Environmental Science, Engineering and Technology

nova
science publishers
www.novapublishers.com

Environmental Science, Engineering and Technology

Atmospheric Aerosols: Properties, Sources and Detection
Binoy K Saikia, PhD (Editor)
2022. ISBN: 979-8-88697-211-5 (Softcover)
2022. ISBN: 979-8-88697-341-90 (eBook)

Applications of AHP Methodology for Decision-Making in Cleaner Production Processes
Fisnik Osmani, PhD and Atanas Kochov, PhD
2022. ISBN: 978-1-68507-882-9 (Softcover)
2022. ISBN: 979-8-88697-012-8 (eBook)

Flow Diagrams Applied to Microalgae-Based Processes
Ihana Aguiar Severo, PhD
2022. ISBN: 978-1-68507-742-6 (eBook)

Ecological Footprints: Management, Reduction and Environmental Impacts
Armano den Hartogh (Editor)
2022. ISBN: 978-1-68507-548-4 (Softcover)
2022. ISBN: 978-1-68507-606-1 (eBook)

Understanding Abiotic Stresses
Krishan K. Verma, Tatiana M. Minkina, PhD, and Vishnu D. Rajput, PhD (Editors)
2022. ISBN: 978-1-68507-508-8 (Hardcover)
2022. ISBN: 978-1-68507-552-1 (eBook)

More information about this series can be found at
https://novapublishers.com/product-category/series/environmental-science-engineering-and-technology/

Seema Nara and Smriti Singh
Editors

Emerging Environmental Applications of Nanozymes

nova
science publishers
www.novapublishers.com

DOI: 10.52305/RFFX4767.

Library of Congress Cataloging-in-Publication Data

ISBN: 979-8-88697-552-9

Published by Nova Science Publishers, Inc. † New York

Contents

Preface

Nanozyme is a class of nanomaterial that possesses intrinsic enzyme activity which mimics the catalytic activity of natural enzymes. Ever since 2007, with the reporting of peroxidase mimic activity in Fe_3O_4 nanoparticles, a plethora of nanomaterials have been reported to possess diverse catalytic activities including peroxidase, oxidase, phosphatase, catalase, haloperoxidase, etc. The research activities on nanozymes have witnessed boom over the past decade due to their high catalytic activity, low cost and high stability. The burgeoning field of nanozymes is still in its infancy as research is underway on various aspects of nanozymes such as their underlying mechanism of action, engineering efficient nanozymes, and exploring their wide applications.

Nanozymes are being used in various spheres of science such as diagnostics, therapeutics, and electronics, however their applicability in different ways to monitor and remediate the environmental pollutants is naïve and bears immense future potential. In recent years, applications of nanozymes are being explored for environmental monitoring. They are being used to develop new methods for sensitive detection of environmental pollutants, removal/extraction of these pollutants from soil or water, and their degradation to relatively safer by products. However, their effective implementation in environmental field to support sustainable growth is yet challenging as it needs deeper understanding of the fundamental principles of their catalytic mechanism, and subsequently design new nanozymes with high selectivity and reusability. In addition, efforts on real time monitoring and their commercial production are also required. These have been the motivation to write this book via assimilation of present and past developments in this emerging field of research. Hence, this book intends to cover the latest trends in usage of nanozymes for environmental health within nine concise chapters.

Chapters 1-4 discusses on the use of nanozymes for detection of hazardous chemicals such as pesticides, organic pollutants, environmental pathogens and food contaminants in environmental or food samples. These chapters present various strategies to detect the aforementioned chemicals using artificial nanozymes. Apart from identification, strategies to remove the contaminants hold significance to get rid of emerging pollutants. In this context, Chapter 5 emphasizes on the ways of using nanozymes for degradation or removal of contaminants like dyes. The book also includes Chapter 6 that discusses futuristic scope of using nanozymes for defense against microbial attacks. The field of maximizing crop production through the use of nanozymes is newly emerging and is discussed in Chapter 7. The last chapter gives a brief overview of the applications of nanozymes in biomedical field as this is also an emerging area where nanozymes are being applied.

This book compiles the existing trends in applying nanozymes for environmental pollutants detection or removal. Certainly, the book will be valuable for students, researchers and scientists of diverse fields wherein the applications of nanozymes could be integrated.

(Dr. Seema Nara)
Lead Editor

Acknowledgments

The editorial team acknowledges the excellent contributions of the authors and co-authors of all chapters who dedicated their worthy time and efforts for their chapters. We especially thank the staff and publishing team of Nova Science Publishers for their support and cooperation.

List of Abbreviations

AA- Ascorbic Acid
ABTS- 2,2-azino-bis 3-ethylbenzothiazoline-6-sulfonic acid
AChE- Acetylcholinesterase
AFM- Atomic force microscopy
AMR- Antimicrobial resistance
AQI- Air Quality Index
ATCh- Acetylthiocholine
ATP- Adenosine Tri Phosphate
BG- Berlin green
CAT- Catalase
CCK 8- Cell Counting Kit-8
CL- Chemiluminescence
CM- Carbamate
CNZ- Copper nanozyme
CuTA- Copper Tannic Acid
DAB- 3,3′-Diaminobenzidine
DAP- 2,3-diaminophenazine
DDT- Dichlorodiphenyltrichloroethane
DMAE- DNase Mimetic Artificial Enzyme
DNA- Deoxy ribo Nucleic Acid
DOPA- 3,4-Dihydroxyphenylalanine
DPV- Differential pulse voltammetry
EAACI- European Academy of Allergology and Clinical Immunology
ECM- Extracellular matrixes
ELISA- Enzyme linked Immuno Sorbent Assay
EPA- Environmental Protection Agency
EPS- Extracellular polymeric substances
ESN- Electronic Serial Number

FARE- Food Allergen Research & Education
FDA- Food and Drug Administration
FTIR- Fourier transform infrared spectroscopy
GC-MS- Gas Chromatography-Mass Spectrometer
GO- Graphene oxide
GO- Graphene oxide
GOX- Glucose Oxidase
GPx- Glutathione peroxidase
HADA- Hyaluronic acid-dopamine
h-BN- Hexagonal boron nitride
HCV- Hepatitis C virus
HIV- Human immunodeficiency virus
HIX- Hybrid ion exchange
HPLC-MS- High Performance Liquid Chromatography-Mass Spectrometer
HQ- Hydroquinone
HRP- Horseradish peroxidase
HUVEC- Human umbilical vein endothelial cells
IARC- International Agency for Research on Cancer
IL- Inter Leukin
KAN- Kanamycin
Kb- Binding constant
LDH- Lactate dehydrogenase
LFIA- Lateral Flow Immunoassay
LMG- Leucomalachite green
LOD- Limit of Detection
MG- Malachite green
M-HFn- Magnetic human ferritin nanozyme
MNPs- Magnetic nanoparticles
MOF- Metal organic framework
MRI- Magnetic resonance imaging
MTS- (3-(4,5-dimethylthiazol2-yl)-5-(3-carboxymethoxyphenyl)-2-(4-sulfophenyl)-2H-tetrazoliumsodium salts
MTT- Methylthiazolyl tetrazolium
MWNT- Multiwalled carbon nanotubes
NG- Graphene doped with nitrogen
NMM- N-methylmorpholine
N-PCNSs- Nitrogen@carbon nanospheres

NSG- Co-doped graphene with nitrogen and sulfur
NTA- Nitrolotriacetic acid
NTPS- Nucleoside triphosphates
OER- Oxygen evolution reaction
OP- Organophosphates
OPD- O-phenylenediamine dihydrochloride
ORR- Oxygen reduction reaction
OTA- Ochratoxin A
OXD- Oxidase
PAA- Polyacrylic Acid
PB- Prussian Blue
PCR- Polymerase chain reaction
PCT- Photo catalytic Therapy
PDT- Photo dynamic therapy
PEG- Polyethylene glycol
POD- Peroxidase
pNP- Para-nitrophenol
POD- Peroxidase
PL- Photoluminescence
PP- Polypropylene
PS- Polystyrene
PTT- Photo thermal therapy
PVP- Polyvinylpyrrolidone
PW- Prussian white
PY- Prussian yellow
rGO- Reduced graphene oxide
RB- Rhodamine
ROS- Reactive oxygen species
RTPCR- Reverse Transcriptase PCR
SDZ- Sulfadiazine
SEM-EDX- Scanning electron microscopy with energy dispersive X-ray spectroscopy
SERS- Surface-enhanced Raman scattering
SOD- Superoxide dismutase
SOM- Soil organic matter
TCs- Tetracyclines
TEM- Transmission electron microscopy

TMB- 3,3,5,5-tetramethylbenzidine
TMO- Transition Metal Oxide
TMD- Transition Metal Dichalogenide Nanosheets
TNF- Tumour necrosis factor
VEGF- Vascular endothelial growth factor
VOCs- Volatile organic compounds
WHO- World Health Organisation
ZIF-8- Zeoliticimidazolate framework-8

Chapter 1

Nanozymes in Food Contaminant Detection

Juri Goswami[1,2], **Lakshi Saikia**[3]
and Parasa Hazarika[1,2]
[1] Jorhat Institute of Science and Technology, Jorhat, Assam, India
[2] Assam Science and Technology University, Jalukbari, Guwahati, Assam, India
[3] Advanced Materials Group, Materials Sciences and Technology Division,
CSIR- North-East Institute of Science and Technology, Jorhat, Assam, India

Abstract

Food safety hazards emerging from exogenic contaminants have been considered a major threat to human health globally. Hence, monitoring food quality is essential for ensuring health safety and maintaining fitness. The evolution of suitable methods to efficiently detect hazardous contaminants in food has attracted enormous attention in various food processing and consumption sectors. Among them, nanozymes are emerging enzyme-mimetic nanomaterials with excellent catalytic properties and high stability which have been innovatively used as efficient tools for the detection of food contaminants.

Herein, nanomaterials, including quantum dots, metal/metal oxide nanoparticles and organic fluorescent-based molecules exhibiting enzymatic (peroxidase, oxidase, and catalase) mimicking properties that act as biosensors of food contaminants are comprehensively summarized. We will also discuss the current understanding of the mechanisms and recent progress of the nanozyme-mimicking sensors used for the detection of harmful contaminants in food samples.

Moreover, this chapter also encompasses the present status of nanozymes explored in the selective and sensitive detection of hazardous food adulterants including pesticide residues, toxic metal ions, antibiotics, pathogens, additives, veterinary drug sediments and hydrogen peroxide.

In: Emerging Environmental Applications of Nanozymes
Editors: Seema Nara and Smriti Singh
ISBN: 979-8-88697-552-9

Finally, the advantages and limitations of nanozyme-based sensors for food contaminants detection will be highlighted to emphasize the current challenges and future trends in industrial applications.

1. Introduction

Communal awareness over food safety and food quality has been gaining massive attention over the past few years. According to World Health Organization (WHO) definition, "Foodborne illnesses are diseases, which are either infectious or toxic in nature, caused by agents that enter the body through the ingestion of contaminated food." Therefore, the monitoring of toxicity in food has been considered a significant way to safeguard food safety. Hence, designing advanced sensing techniques for the rapid detection of contaminants in food is of utmost importance. However, consumption of food has been shifting from traditional nutrient-based to packaged products that are rich in several additives. These additives cause massive risk to human health.

With the increasing demand for food to feed the huge population, there has also been a rise in food contamination that raises the possibility of infiltrating contaminants at every stage of the production and transportation processes (Soon et al., 2020; Manning et al., 2019; Li et al., 2018; Liu et al., 2018). Contamination of food with toxic substances could be occurred at any stage i.e., growing, harvesting, processing, storing, shipping or preparing. Similarly, cross-contamination may also lead to food toxicity through the transference of harmful microorganisms from one surface to another. Contaminants such as thiometon, antibiotics, pesticides, mycotoxins, heavy metals, food allergens and viruses were found in several foods at concentrations exceeding their maximum limits permittable by World Health Organization. Food toxicity may also arise from the direct consumption of uncooked raw vegetables, salads and seeds; hence, harmful microorganisms could not be destroyed in the process of cooking which might lead to serious food poisoning and illness.

Various laws and limitations have been established by World Health Organization to elevate nutrition in food in order to satisfy public needs and to protect humans from infection with toxic ingredients. The World Health Organization also aims to encourage worldwide prevention, detection, and immediate response to public health threats associated with hazardous food products. Achievement of trust and confidence in the safe food supply to the consumers have also been ensured by WHO. As a result, conducting an efficient, trustworthy and sensitive analytical technique for screening contaminants in food matrices is extremely desirable. Generally, various detection techniques like liquid chromatography-mass spectrometer (HPLC-MS), gas chromatography-mass spectrometer (GC-MS), high-performance liquid chromatography (HPLC), surface-enhanced Raman spectroscopy and many others were well implemented for the detection of harmful contaminants in food (Hunter et al., 2010; Yoshioka et al., 2008; Fu et al., 2016; Zhou et al., 2019; Fu et al., 2019; Luo et al., 2016; Bajkcz et al., 2020). Despite the excellent sensitivity and accuracy of the results produced by these analytical techniques, a number of drawbacks are also present, such as the need for time-consuming procedures, complicated machinery and reliance on high-end equipment operated by skilled personnel. In contrast, fluorometric, colorimetric, ratiometric, and electrochemical methods have been extensively identified as promising tools to effectively detect food contaminants due to their tremendous sensitivity, rapid detection with easily readable results and handy instrumentation (Pereira et al., 2018; Martins et al., 2017).

In past times, natural enzymes were extensively employed in the agri-food sector for processing food products. However, natural enzymes have drawbacks such as low thermostability and a small pH window, which cause the enzymes to denature and significantly diminish their enzymatic activity. Susceptible denaturation complicates the analysis of sensing and monitoring outputs, which could lead to a false positive or negative conclusion. However, nanozymes compared to natural enzymes have several advantages like low cost, easy synthesis procedure, superior resistivity to severe environmental conditions, long-term stability, and on-demand tunable properties (Wang et al., 2018; Zhang et al., 2019; Lv et al. 2018). Additionally, nanozymes mimic natural enzymes and display enzyme-like properties. Therefore, the huge attention of researchers has been shifting towards the production of inorganic nanomaterials with enzyme-mimicking properties possessing higher thermal and chemical stability. Nanozymes are also known to have utility in numerous fields like chemosensing, biosensing, environmental care, agri-food monitoring and therapeutics agents (Huang et al., 2019; Wang et al., 2016; Gao et al., 2020; Wong et al., 2021). Recently, a great number of nanozymes with diverse structures and compositions such as carbon quantum dots, nano-porphyrin, metal/metal oxide nanocomposites, organic fluorescent molecules and aptamers have been broadly applied in food safety analysis with rapid detection, easy instrumentation, excellent specificity and stability.

Hence, in this chapter, current advances in nanozyme-built sensors for selective sensing of various food contaminants have been discussed. A brief introduction to nanozymes with a focus on their varieties and elaboration of nanozyme-enabled analytical techniques are also explored. By modeling the nanozymes with surface functionalization, detection of endogenous and exogenous contaminants in food products through these powerful techniques has been reviewed. Furthermore, a series of applications of the nanozyme-based sensors for the detection of (pathogens, viruses, heavy metals, toxins, allergens, pesticides, herbicides, antibiotics, etc.) in food has been elaborately analyzed, and possible outlook towards the future development of nanozymes is finally proposed.

2. Types of Enzyme Mimics

Determined by the boundless investigations of researchers, several forms of inorganic nanomaterials possess inherent enzyme-mimic activities, thus providing a bridge between nanotechnology and biology. Natural enzymes possess inherent catalytic activity through a sole active site to catalyze a definite chemical reaction but nanozymes can have numerous active sites depending on their route of synthesis to increase their catalytic properties. This allows them to selectively conjugate and bind with substrate molecules. Nanozymes can be categorized into four different classes. Type 1 nanozymes have been considered active metal center mimic that comprises transition metal oxides (Fe_3O_4, CeO_2) (Gao et al., 2007; Palizban et al., 2010). Metal nanoparticles and 0D/2D/3D carbon nanomaterials have been classified into class II which is mainly functional mimics (Lin et al., 2018; He et al., 2013). Type III are hybrids of type I and type II and metal-organic frameworks (MOFs) (Li et al., 2019). Single-atom nanozymes and bi-metallic alloys with etched channels have been classed as 3D structural mimics (type IV) (Huang et al., 2019; Benedetti et al., 2018).

These inorganic nanomaterials mimic a variety of natural enzymes, like horseradish peroxidase (POD/HRP), superoxide dismutase (SOD), catalase (CAT), oxidase (OXD), glucose oxidase (GOX), phosphatase, etc. and are known to react with different chromogenic substrates (Wua et al., 2020; Heckert et al., 2008; Fan et al., 2011; Asati et al., 2009; Fan et al., 2018; Wang et al., 2018; Liu et al., 2013; Li et al., 2017). Chromogenic substrates are of various types including 3,3,5,5-tetramethylbenzidine (TMB), 2,2-azino-bis (3-ethylbenzothiazoline-6-sulfonic acid) (ABTS), o-phenylenediamine dihydro-chloride (OPD) etc. For example, peroxidases mimic in the presence H_2O_2 react with TMB, ABTS, OPD and catalyze their oxidation to oxidized products. For example, magnetic nanoparticles like Fe_3O_4 mimic POD to transform the substrate TMB into an oxidized product (TMB-ox) in presence of hydrogen peroxide thereby forming a blue color solution after incubating for a specific time (Vallabani et al., 2017; Fan et al., 2017). These enzyme mimics have been extensively applied in the detection of many analytes like glucose, and H_2O_2. Compared with peroxidase mimics, much research work has been recently developed to synthesize nanozymes with OXD mimics as it acts as a catalyst to carry out redox reactions in the absence of unstable H_2O_2. N-doped $Fe_3C@C$ composites act as dual mimics. A) oxidase-mimics to catalyze the oxidation of TMB by molecular oxygen, releasing H_2O, B) even if O_2 is transformed into H_2O_2, it further reduces to H_2O employing the peroxidase-mimic behavior of N-doped $Fe_3C@C$ nanozymes (Yang et al., 2017).

Pirmohamed et al., (2010) reported nanoceria performing as CAT mimics for catalyzing the decomposition of H_2O_2. CAT is an enzyme found in all biological organisms that defends the cells from exposure to H_2O_2. GOD-like activity, which is widely implanted for the oxidation of glucose to produce H_2O_2 has captured substantial attention in biological science and therapeutics. Luo et al., (2010) exhibited GOX-like activity of synthesized gold nanoparticles (AuNPs) that catalyzes the decomposition of glucose into gluconic acid, moreover, using a chromogenic indicator (ABTS), glucose detection could be easily performed. Recent studies revealed that there could be a possibility of existing multiple catalytic activities of a single nanozyme having numerous active sites, influencing the spatial arrangements of diverse atoms to carry out redox reactions in the electron-transfer process.

Another important class of iron-based nanozyme is Prussian blue (PB), which is interchangeably linked as Fe^{2+}-C≡N-Fe^{3+} bonds thereby forming cubic cyanide bridge systems (Farka et al., 2018). The tunable energy levels and redox potentials link them as excellent electron transferring groups thus mimicking multienzyme activities like POD, CAT, SOD, etc. Characteristically, PB nanozymes when oxidize get transformed either to Berlin green (BG) or to Prussian yellow (PY) and on reduction transform to Prussian white (PW). In acidic media, hydrogen peroxide oxidizes PB to BG/PY (E=1.40 V), acting as peroxidase mimics thereby oxidizing TMB with the aid of H_2O_2. In an alkaline solution, PB displays intrinsic catalase-like activity (Zhang et al., 2016).

There exist a vast range of nanozymes, which are much more operative in comparison to the natural enzyme and have been known to have real application in multiple scientific fields. In recent years, numerous nanozymes offer flourishing applicability in the agri-food industry and have been extensively implanted for the detection of toxic ingredients in food. Generally, the nanozyme-built detection techniques could be achieved through a suitable method where the target analytes and nanozymes are allowed to react. The results obtained from the analysis are mainly achieved from different analytical assays like colorimetric, electrochemical, fluorescent, and chemiluminescent signals.

3. Catalytic Mechanisms

For a creative design of a nanozyme with improved catalytic performance, understanding its reaction mechanisms is a crucial factor. Their activities are greatly dependent on several factors like size/shape, surface modifications, surface molecular imprinting effect, etc.

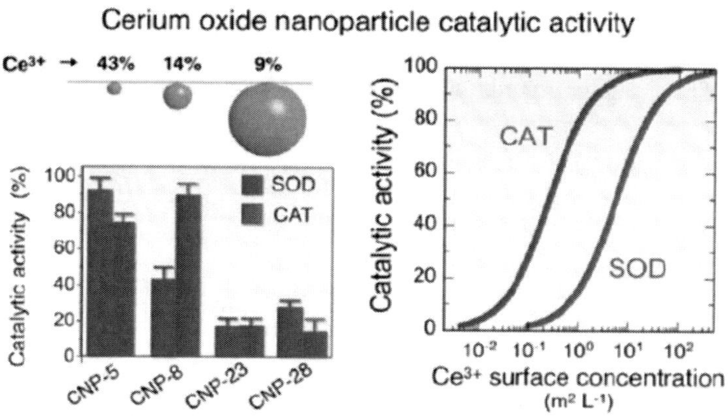

Reprinted with permission from Dong et al., 2019. Copyright © 2019, American Chemical Society.

Figure 1. Cerium oxide nanoparticle with size-dependent catalytic activity.

3.1. Size and Shape Dependence

The size and shape of nanozymes are some significant criteria that can affect their mechanistic activity. As the surface of nanozyme is constituted of diversified catalytically active particles, therefore excellent catalytic activities will be achieved from small diameter particles constituting the nanozyme

(Dong et al., 2019). For example, for performing a special redox cycle between Ce^{3+} and Ce^{4+}, cerium oxide nanoparticles (CeO_2) exhibit either SOD or CAT mimic properties depending on the mechanism of electron transfer (Baldim et al., 2018). As the size of CeO_2 decreases, more Ce^{3+} ions constitute the surface. Thus, smaller CeO_2 offers a greater catalytic activity in comparison to that of larger CeO_2 (Figure 1).

The size of nanozymes can also be related to the coordination number of the atoms on the surface of the catalyst, thereby motivating the catalytic activity (Cao et al., 2016). The lower coordination and hence smaller CeO_2 strongly binds to the reactants, intermediate, or product molecules with an excellent binding constant (K_b). The shape-dependent catalytic activity with varying specific crystal planes could be attained by following a selective synthesis procedure for nanozymes. Ghosh et al., (2018) synthesized V_2O_5 nanoparticles via different routes to achieve four varying morphologies in order to imitate the activity of glutathione peroxidase (GPx) mimic. The as-obtained morphologies include (VNw), (VSh), (VNf), and (VSp) named for Vanadium nanowires, nanosheets, nanoflowers and nanospheres respectively. The steady-state kinetic parameters were studied for all the morphologies. The maximum velocity (V_{max}) attained observed was very much different from each other i.e., VSp > VNf > VSh > VNw (Figure 2).

Figure 2. (a-d) SEM and TEM photographs of (a) VNw, (b) VSh, (c) VNf, and (d) VSp.

3.2. Surface Modifications

To build the required activities of the nanozyme, proper surface modification or functionalization is a dynamic and effective method. Thus, the surface modification could be achieved by regulating different parameters like charge, pH, active sites, substrate adsorption/desorption and so on. For example, the adsorption of F^- or SO_4^{2-} on the surface of CeO_2 offers a surface charge from positive to negative thereby increasing the binding affinity for TMB and hence increasing the enzymatic activity of CeO_2. Moreover, the adsorption of F^- on the surface of nanozyme modifies the surface state of atoms, thereby declining the formation of oxidized products and encouraging desorbed products. Furthermore, SO_4^{2-} have powerful electron-withdrawing behavior which encourages molecular oxygen adsorption on the CeO_2 surface. Therefore, the redox cycle reaction ($Ce^{3+} = Ce^{4+} + e^-$) is facilitated by the adsorption of O_2 on CeO_2, thereby uplifting the binding affinity between O_2 and substrates (Huang et al., 2017; Liu et al., 2016).

3.3. Small Molecule Modification

Catalytic performances of nanozymes can be increased by the addition of various amino acids, nucleoside triphosphates (NTPs) and many others to its active sites. Specifically, several chiral amino acids could also modify the catalytic properties of nanozymes, consequently optimizing their stereo-selectivity. Stereospecificity is very much crucial in the biomedical field as it influences the nanozyme to react with a specific receptor or target molecule after confronting them in the body.

Sun et al. (2017) designed chiral CeNPs mimicking OXD by modifying the catalytic sites with D or L Phe (phenylalanine). The substrate 3,4-Dihydroxyphenylalanine (DOPA) enantiomer can be oxidized by CeNPs thereby exhibiting an absorbance peak at 475 nm. On measuring the UV-vis absorbance, it can be seen that CeNPs alone didn't show any stereoselectivity for DOPA. However, modifying CeNPs with D-Phe (CeNP@D-Phe) increases the selectivity to oxidize L-DOPA, while CeNPs modified with L-Phe (CeNP@L-Phe) show high selectivity toward D-DOPA. The difference in catalytic behavior is proportional to the hydrogen bonds created through the reaction of L or D DOPA enantiomers with L or D Phe enantiomers.

3.4. Macromolecular Coating

Modification of catalytic surface could also be achieved through the interaction of some polymers, like PP (polypropylene), PVP (Polyvinyl-pyrrolidone), PAA (Polyacrylic Acid), PEG (Polyethylene glycol), and PS (Polystyrene) which has been known to be a common interest of some researchers. The addition of biopolymers to the nanozyme could be achieved via several interactions like electrostatic, coordination and hydrogen bonding. Moreover, macromolecular modification improves the colloidal stability and the affinity of nanozyme to substrates thereby increasing its catalytic efficiency (Hizir et al., 2016).

Liu et al., (2015) worked on the enzymatic activity of CeO_2 by surface modification with DNA. CeO_2 in interaction with DNA activates its POD behavior. On the other hand, the OXD mimic activity of CeO_2 declined on conjugation with DNA. Apparently, the twin-mimicking behavior might be probably due to the different reaction mechanisms for OXD and POD mimics exhibited by CeO_2. When CeO_2 acts as an OXD mimic, the particles on the surface act as an oxidant and for initiating the catalytic reaction the substrates bonded to the particles of the catalyst. However, the POD mechanism is completely different. Here, the free radicals produced from H_2O_2 act as oxidants thereby oxidizing the substrates. Hence, the DNA exhibited the surface blocking effect which creates inhibition of the OXD behavior of CeO_2 but has a negligible effect on the POD behavior.

3.5. Surface Molecular Imprinting

Surface molecular imprinting technology is an emerging technique that engineers the substrate binding sites for increasing the substrate specificity of nanozymes (Zhang et al., 2017). For example, POD enzyme mimic Fe_3O_4 nanoparticles in presence of H_2O_2 and can bind to different chromogenic substrates (Figure 3). However, the use of monomers such as neutral acrylamide and N-isopropylacrylamide with impregnated TMB or ABTS template molecules could uplift the binding specificity of Fe_3O_4 to the imprinted molecules (Zhang et al., 2017).

Natural enzymes could be categorized into oxidoreductase which is classified as OXD, POD, SOD and hydrolases. These enzyme mimics follow a redox reaction in which hydrogens, oxygen atoms, or electrons are transferred between atoms of a molecule. The redox reactions are followed by

POD, SOD and OXD types whereas hydrolases include enzymes such as peptidase, phosphatase, nuclease etc. for hydrolyzing a chemical reaction. The pH of the reaction medium has a major effect on the catalytic activities of nanozymes thus exhibiting intrinsic pH-switchable enzymatic activities. The pH-dependent nature of the nanozymes arises from the adsorption of hydrogen peroxide on the surface of the catalyst and its decomposition is basically an acid-base type of catalysis involving OH^- and H^+ under different pH conditions (Li et al., 2015).

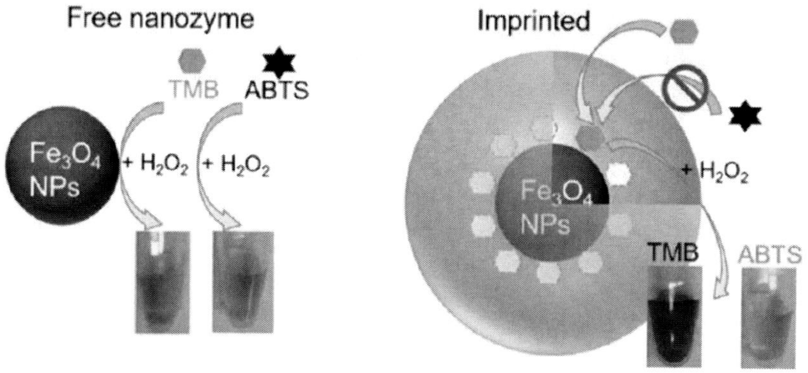

Figure 3. Schematic diagram of Fe_3O_4 NPs in comparison to free and impregnated nanozyme.

In an acidic medium, H_2O_2 exhibits a base-like decomposition mechanism to display POD mimic behavior, while in a basic medium, CAT mimic activity is exhibited following an acid-like decomposition pathway. In acidic conditions, the pre-adsorbed H triggers the decomposition of H_2O_2. Hydrogen peroxide decomposes to active species OH^* through the breakage of O-O bond. Here, OH^* dissociates into reactive oxygen species forming H_2O and O^*(reactive oxidizing species). The reactive oxidizing species O* withdraw H atoms from TMB to manifest the nanozyme for exhibiting POD behavior. In a basic medium, the pre-adsorbed OH on nanozyme surface stimulates H_2O_2 towards acid-like decomposition. H_2O_2* reacts with OH* following the formation of unstable species (H* and HO_2*). The reactive H* reacts with OH* to produce H_2O*, whereas HO_2* drifts H a different H_2O_2*, residual O_2* thereby forming H_2O_2*, H_2O* and OH*. This mode of the reaction mechanism is followed by CAT-like nanozyme.

The catalytic mechanism for oxidase-like behavior initially starts through the dissociative adsorption of oxygen on nanozyme surface. Here, the metal active centers trigger strong adsorption of O_2 on the surface thereby weakening the O-O bond which gives rise to O-O bond elongation with zero bond order forming O*. The hydrogen atoms available from surrounding organic substrates (TMB) present in the reaction medium can be abstracted by these O^* atoms (Shen et al., 2015). Dong et al., (2014) in Figure 4 reported the enzymatic mimics of (Cobalt oxide) Co_3O_4 in different pH mediums. Co_3O_4 possesses pH-dependent activities with multienzyme-like behavior because of the superior redox potential values for Co^{3+}/Co^{2+}(1.3 V), as displayed below.

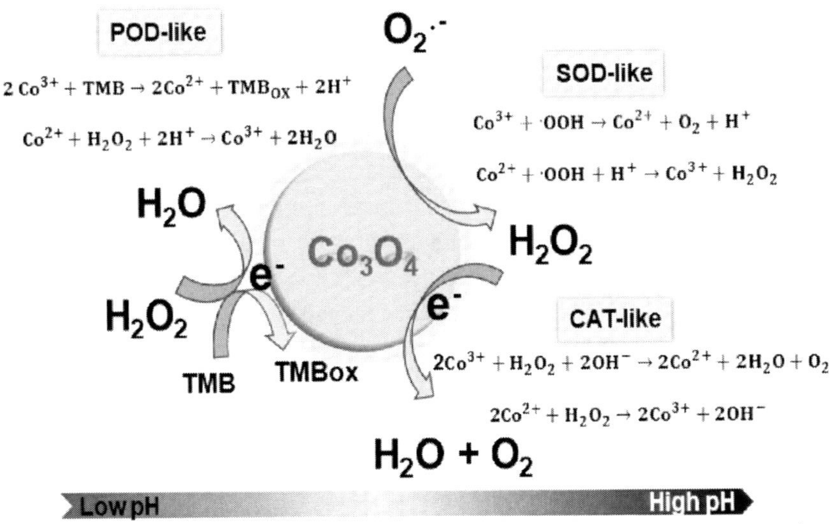

Figure 4. POD, SOD, CAT like activities exhibited by Co_3O_4 nanozymes.

4. Signal Amplification of Nanozymes-Based Sensors

So far, the nanozymes-based sensors have been applied in various applications based on the mode of mediated signal amplification (e.g., colorimetric, fluorescent, chemiluminescent, electrochemical, ELISA etc.). Herein, we discuss briefly different modes of signal amplification of nanozymes-based sensors just to accumulate knowledge of various analytical techniques.

4.1. Colorimetric Sensors

This type of sensing is mainly performed under UV-vis spectrometry. Here, the sensing is mainly achieved on the basis of the change of color of the reaction solution and absorbance value with the addition of different concentrations of analytes. Colorimetric sensors have attracted wide attention owing to the change of color of the reaction solution which provides easily readable results and fast visual detection through the naked eye. This type of sensing is very much suitable for fast detection as it needs portable equipment which is handy and does not require trained personnel to handle the instrument. Nanozymes undergo catalytic oxidation with various substrates to produce colorimetric output signals (Yu et al., 2019).

4.2. Fluorometric Sensors

Fluorescence sensors are created mainly based on the fluorescence intensity enhancement ("turn-on") or quenching ("turn-off") of the sensor probe by a quencher. To date, nanozymes-based fluorescence sensors have been widely applied for the generation of fluorescence signals and intensively utilized in biomedical, pathogenic microorganisms' assay, and pollutant monitoring processes. Nanozymes-based fluorescence sensors achieved the advantages of excellent selectivity, superb sensitivity, simple instrumental operation and short response times (Altunbas et al., 2020).

4.3. Chemiluminescent Sensor

Chemiluminescent sensing is a promising analytical tool that has been created with reference to the emission of light from a chemical reaction. This light emission-mediated technique has been implemented for the determination of numerous target analytes due to its notable advantages of simplified operation, modest equipment, small LOD, and varied linear range (Abhijith et al., 2013).

4.4. Electrochemical Biosensors

This type of sensing is mainly constructed based on the difference in output signals as a result of chemical transformations in the reaction of the target analytes and electrode-immobilized recognition elements. The concentrations of target analytes have a great relationship to the generation of electrical signals, which can be applied for qualitative and quantitative detection of target molecules. To date, electrochemical sensors have found wide application in multiple fields, such as biological, environmental care, analytes determination, and clinical diagnosis due to their modest operation, low cost, amazing stability, and short response time. To further improve the sensitivity of the electrochemical sensors, the catalyst has to be properly selected to modify the electrode with even dispersion and enhanced electrocatalytic sites (Khairy et al., 2018).

5. Nanozymes in Food Contaminant Detection

5.1. Detection of Pesticides and Herbicides

Pesticides, a class of chemical compounds that are widely used for agricultural purposes to regulate pest management and plant growth have been known as a rising threat to the environment (Zhang et al., 2016; Pope (1992); Aragay et al., 2012). To ensure better health care, the determination of harmful pesticides in food products, water and the surrounding environment is very much essential. Zhu et al., (2020) constructed a colorimetric sensor based on doping in graphene by heteroatoms to form (graphene doped with nitrogen (NG), co-doped graphene with nitrogen and sulfur (NSG), and graphene oxide (GO)) for sensing pesticides like lactofen, fluoroxypyr-meptyl, bensulfuron-methyl, fomesafen, and diafenthiuron.

The advantage of this sensor lies in the fact that the heteroatom doping exhibits different peroxidase activities depending on the nature of the element doped. The active sites for reaction could be selectively masked by the adsorption of different pesticides on the surface of the nanozyme. On the basis of this principle, five pesticides with concentrations varying in the range of 5 to 500 μM could be easily detected (Figure 5).

Similarly, Wei et al., (2019) worked on CeO_2 nanozymes with dual-mode functions for the detection of organophosphorus pesticides. CeO_2 nanozymes exhibit phosphatase mimic activity for hydrolyzing organophosphorus pesticides to produce para-nitrophenol (p-NP). The reaction scheme depends on the color generation where p-NP unveiled a bright yellow color through a change in the color intensity with varying pesticides concentration. Thus, through monitoring the color change of the reaction product and evaluating the change of absorption peak at 400 nm, an effective smartphone-based colorimetric technique for methyl paraoxon detection was created. Furthermore, the above method has been successfully applied in real samples for the determination of pesticide residue.

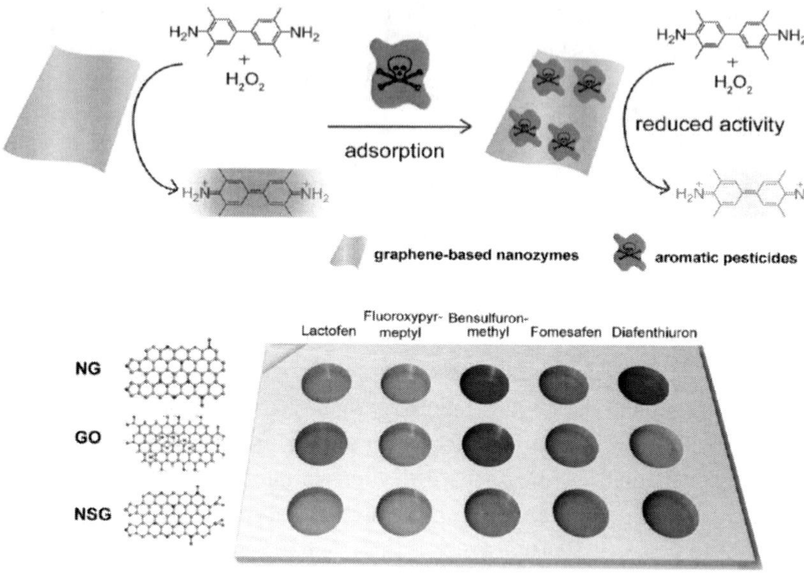

Figure 5. Schematic diagram based on heteroatom-doped graphene nanozyme sensor arrays for detecting aromatic pesticides.

The detection of herbicides is also a challenging topic in agricultural science. Luo et al., 2021 demonstrated peroxidase mimic activity of Co_3O_4 nanoplates on a polyester fiber membrane as an effective method for detecting glyphosate herbicide in agricultural crops. On addition of herbicide (glyphosate) to the solution, the peroxidase-mimicking activity of porous Co_3O_4 nanoplates could be openly inhibited, thereby the color transformation from blue to transparent proved to be a visual detection technique for glyphosate detection. The prepared nanozyme exhibits excellent selectivity towards glyphosate detection with a LOD of 0.175 $mg\cdot kg^{-1}$. Wu et al., (2021) constructed 2D MnO_2 nanosheets (MnNS)) with both OXD/POD behavior, hence an electrochemical detector was developed for organophosphate pesticides (OPs) detection without adding H_2O_2. Here, dissolved O_2 acts as a co-reactant and color change could not be observed. Owing to the unique catalytic activity of MnNS arising from the larger surface area, a significant amount of TMB is catalytically oxidized, where a declination of differential pulse voltammetry (DPV) current could be observed. Clearly, MnNS exhibited better sensitivity to thiocholine (TCh) during the catalytic reaction of acetylthiocholine (ATCh) by acetylcholinesterase (AChE). Thus, OP detection by MnO_2 nanosheets based on the reducing activity of AChE was carried out for a LOD value of 0.025 ng mL^{-1}.

Similarly, an excellent sensing tool for OPs could also be performed based on a paper strip sensor. Based on this concept, Huang et al., (2019) synthesized a paper-strip colorimetric sensor for (OPs) sensing. Here, y-MnOOH mimics OXD and acetylcholinesterase (AChE) constituted the detection system where TMB was oxidized in the presence of O_2. The AChE could catalyze acetylthiochlorine (ATCh) to form thiocholine (TCh). The OXD behavior of y-MnOOH decreases as a function of obtained TCh thereby simultaneously reducing y-MnOOH to Mn^{2+}. In this way, a visual and rapid determination of AChE could also be observed with a LOD of 0.007 mU mL^{-1}. This assay was employed for determining dichlorvos pesticides with LOD of 0.14 ng mL^{-1} with potential application in real sample (Chinese cabbage) analysis.

5.2. Detection of Food Mycotoxins

Mycotoxins mainly originated from plants and harmful microbes cause life-threatening symptoms like nerve intoxication, dehydration in the body, and even brain failure if consumed above permittable limits (determined by WHO). Mycotoxins can have an impact on the safety and quality of agricultural goods, along with the accompanying processed foods, feeds, and animals.

A frequently distinguished mycotoxin in food products named as Aflatoxin B_1 (AFB1) falls in the list of most toxic mycotoxins detected in recent years (Xie et al., 2017). International Agency for Research on Cancer categorized AFB_1 as a class 1 carcinogen (Ostry et al., 2017). The maximum allowable limits for AFB_1 in foodstuffs are (1.0-20 ng/g) (Alshannaq et al., 2017; Rocha et al., 2014; Egmond et al., 2007). In past years, several analytical methods such as HPLC, thin-layer chromatography and ELISA have been applied for the effective detection of AFB_1 in foodstuffs. Though these methods display good accuracy, still, they suffered from serious disadvantages regarding the speed of the detection process (Turner et al., 2009; Yang et al., 2020). To subjugate these limitations, Xu et al., (2021) worked on MOF-linked (MIL-88) immunosorbent assay for AFB_1 detection in grain drinks by assembling a corresponding antibody Ab2. In this manner, the combined peroxidase-like properties of the MOF and Ab2 could efficiently bind to AFB_1 with a limit of detection of 0.009 ng mL^{-1}. Moreover, the sensing assay was also utilized for Aflatoxin detection in various milk samples.

Lu et al., (2021) worked on PBNPs (Fe^{2+}-C≡N-Fe^{3+}) generated in-situ on magnetic nanoparticles (MNPs) surface to form PBNPs-supported MNPs. The in-situ formation of PBNPs on the MNP surface occurred by the reaction of HCl and $K_4Fe(CN)_6$, confirmed by the UV-vis spectrophotometer where an absorption peak centered at 781 nm was generated accompanied by the color change of the reactant solution. The colorimetric detection of AFB_1 was evaluated as a result of peroxidase mimic activity of PBNPs@MNPs. The absorbance of nanozyme at 652 nm has a positive correlation with the $ABFB_1$ concentration. The color of the oxidized TMB solution was intensified within a definite concentration range with a LOD value of 43.76 $fgmL^{-1}$. Moreover, the nanozyme was also utilized in the fluorometric detection of AFB_1. On the surface of MNP, a hybrid complex was formed through the conjugation of 5′-Cy5-labeled cDNA and aptamer via the linkages of biotin-streptavidin. In the same condition, PBNPs were generated in situ on the surface of the hybrid complex.

Throughout the etching process of MNPs, the formation of Apt/cDNA hybrid complex declined which was established by the recovered fluorescence signal arising due to discharged Cy5, verifying the possibility of fluorescent signal readout facilitated by the generated PBNPs (Figure 6). The fluorescence intensity at various optimal conditions was screened in order to evaluate the discharging behavior of Cy5 encouraged from in situ generated PBNPs on MNPs surface. Hence, the fluorescence quenching by MNPs facilitated the ultrasensitive detection of AFB_1. The fluorescence signal of Cy5 increased steadily with proportion to AFB_1 concentration thereby obtaining a LOD value of 0.54 fg/mL which is much more sensitive than that of the colorimetric technique discussed above.

Another class of mycotoxin known as Ochratoxin A (OTA) is a toxic secondary metabolite with high chemical and thermal stability mainly obtained from Aspergillus and Penicillium species (Pagkali et al., 2018). Consumption of food (wheat, corn, coffee, cocoa, beer, wine and dried fruit) contaminated with OTA resulted in hepatotoxic and renal toxicity illness and is a leading cause of cancer that is declared by the International Agency for Research on Cancer (IARC) (Singh et al., 2020; Zhou et al., 2020).

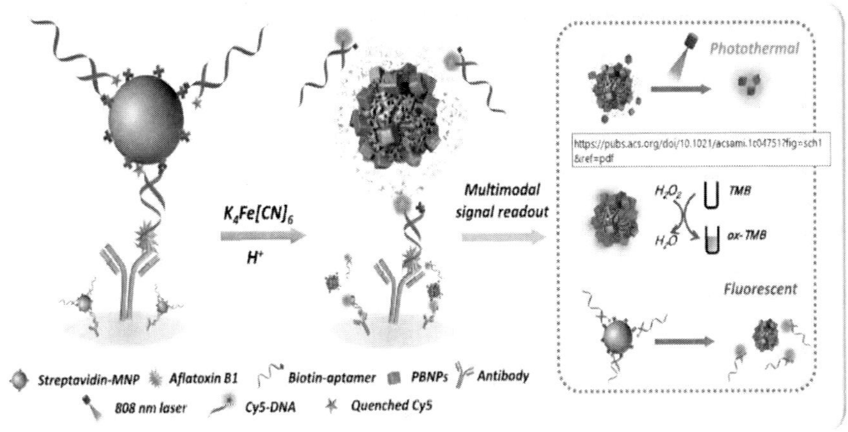

Reprinted with permission from Lu et al., 2021. Copyright © 2021, American Chemical Society.

Figure 6. Schematic diagram based on PBNPs generated in situ for sensing AFB_1.

Moreover, OTA is much more problematic to remove from the ecosystem, hence, it enters the food chain thereby affecting the overall ecosystem. Thus, the construction of suitable techniques for the effective detection of OTA gained the highest priority for the researchers. To date, many traditional techniques like HPLC, HPLC-MS, UPLC-MS/MS, and others have been reported to detect OTA, but these techniques suffer from several disadvantages which are repeatedly discussed above (Yu et al., 2019; Andrade et al., 2017; Zhang et al., 2019; Medina et al., 2021; Alsharif et al., 2019). Therefore, many research fellows developed various detection techniques having high sensitivity and selectivity for OTA sensing in order to monitor food safety.

Huang et al., (2018) constructed an aptamer-based $MnCo_2O_4$ nanozymes for colorimetric sensing of OTA in food products. Here, the $MnCo_2O_4$ nanozyme acts as an OXD mimic to catalyze colorless TMB into an oxidized product with the aid of O_2. The obtained nanozyme could efficiently recognize OTA via assembling a target-specific aptamer. The oxidase-mimic activities of the $MnCo_2O_4$ nanozyme could be selectively used as a probe to detect OTA in maize samples. Similarly, Zhang et al. (2021) demonstrated a cascade reaction for OTA detection based on MnO_2 nanozyme. Here, MnO_2 behaves as an oxidase mimic which can convert TMB to blue ox-TMB. In a colorimetric assay, the blue color of ox-TMB solution decreases with an increase in OTA concentration. Hence, this cascade colorimetric OTA detection demonstrates an extraordinary sensitivity with a LOD value of 0.069 nM and can be employed as a potential sensor in food safety assays.

5.3. Detection of Foodborne Pathogens

Pathogens of various types like Coliforms, Salmonella, Bacillus, Listeria monocytogenes, Cyclospora cayetanensis, norovirus, etc., attack food items leading to various physio and biochemical changes that cause a poisonous effect in food (Wei et al., 2018).

Cheng et al., (2017) constructed an immunoassay-integrated smartphone platform as an efficient nanozyme for pathogen (Salmonella, E. Coli O157:H7) detection. Here, palladium@platinum (Pd@Pt) nanozymes exhibit intrinsic POD activities for catalyzing the TMB oxidation to a blue ox-TMB solution with H_2O_2 assistance. The advantages associated with this immunoassay and smartphone-constituted device lie in the fact that it efficiently eliminates the cross-interference and quickly detects pathogens. The above-mentioned technique also demonstrated the sensing of bacterial pathogens in food products.

Similarly, Xue et al., (2021) designed MnO_2 nanoflowers (MnO_2 NFs) as a POD nanozyme for the detection of pathogens (Salmonella). Here, a lab-on-a-chi biosensor, microfluidic chip and MnO_2 nanozymes were utilized for performing the bacterial detection experiment. Herein, a sandwich complex composed of immune magnetic nanoparticles, salmonella, and MnO_2 NFs was formed using a convergence-divergence spiral micromixer (Figure 7).

On adding TMB to the sandwich complex, a blue-colored solution containing ox-TMB was formed confirming its peroxidase mimic activity. Finally, a smartphone app detection system for bacteria was developed to measure the number of poisonous bacteria detected as a function of CFU mL^{-1}. Thus, this sensing system could easily detect Salmonella with a LOD value of 44 CFU mL^{-1} and has been applied for bacterial detection in chicken products.

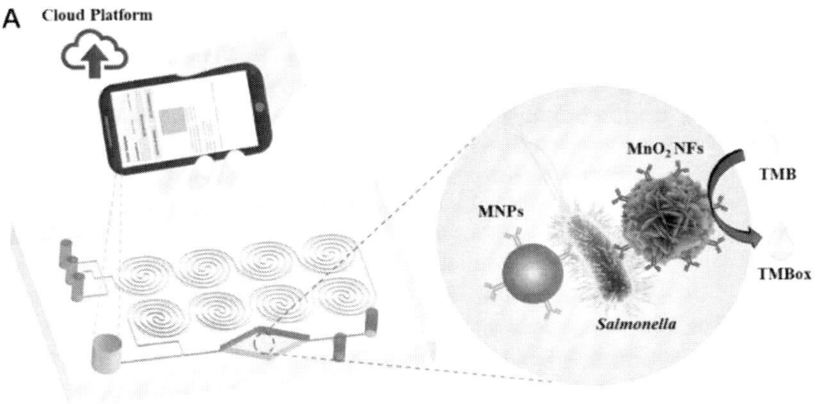

Reprinted with permission from Xue et al., 2021. Copyright © 2021, American Chemical Society.

Figure 7. Schematic diagram for colorimetric detection of Salmonella.

Tarokh et al., (2021) worked on sensitive sensing of Salmonella typhimurium in food based on a colorimetric aptasensor composed of g-C_3N_4@Cu_2O composites. The nanozymes serve a dual-purpose contributing excellent POD-mimics through interconnection to a label-free aptamer. In the presence of TMB and H_2O_2, g-C_3N_4@Cu_2O successfully develops a visible blue color. But the catalytic activity of the composite drastically declines in connection with aptamers. The aptamers bound to their specific target in the existence of S. typhimurium. Then, the POD behavior of g-C_3N_4@Cu_2O was increased by increasing the concentration of S. typhimurium. Moreover, the apt sensor unveiled an outstanding sensing ability in a concentration range (1.5×10^1 to 1.5×10^5 CFU ml^{-1}), with a LOD value of 15 $CFUml^{-1}$ (CFU=colony forming units). The detection system was applied in milk samples thus exhibiting excellent recovery percentage.

Consumer consciousness of health and welfare is growing intensely. Recently, the spread of virus-derived infectious diseases has been growing faster way. Therefore, an efficient sensor for the sensitive detection of viruses in food samples and water bodies is highly required (Chung et al., 2015; Dong et al., 2015). NoV infections are considered as serious infections in food, especially during the winter season. In some countries, like Japan, England, US, China, India and Russia, norovirus (NoV) infection causes serious health hazards (Derrick et al., 2021; Todd et al., 2007; Estes et al., 2019; Wu et al., 2018). Several illnesses like acute gastroenteritis are the root cause of NoV infection. NoV infection mainly happens in the gut, especially in the epithelium of the human small intestine. Diarrhea, vomiting, fever, headaches and abdominal pain may be the possible symptoms of virus infection. NoV is normally contaminated in food and water, therefore even a dose of 100 virus particles/mL can infect a person causing serious health defects (Lane et al., 2019).

Khoris et al., (2019) worked on (Au/Ag NPs) exhibiting POD behavior for colorimetric determination of NoV. The Ag-Au core-shell nanoparticle was formed through the reaction of the Au nanoparticle with $AgNO_3$ in presence of hydroquinone. In presence of H_2O_2/TMB solution, the nanozyme developed intense blue color which is due to the liberation of Ag^+ ions from the Au/Ag NPs surface, thus enhancing the TMB oxidation. The TMB oxidation reaction was stopped which could be observed from a transformation of blue color to yellow by the addition of H_2SO_4 solution. The experimental results confirmed that the nanozyme specifically binds to NoV, and other viruses or proteins didn't show any specificity to the nanozyme. The selective detection by this colorimetric assay was examined by monitoring the change in color of the solution on changing the concentration of clinically isolated NoV using NoV-like particles, with a detection limit of 10.8 pg/mL.

Weerathunge et al., (2019) synthesized gold nanoparticles (GNPs) with POD behavior for colorimetric detection of infective murine NoV. Here, the gold nanoparticles interacted with the target specified MNV aptamer followed by oxidation of TMB with increasing concentration of virus. The efficiency of the nanozyme was further tested in some other microorganisms, human serum and shellfish homogenate demonstrating the high potentiality of the aptasensor to detect NoV in a complex matrix. The most attractive property of this nanozyme was its ultimate fast detection time (10 minutes) and also presenting its ease of applicability and eliminating the need for complicated laboratory infrastructure.

5.4. Detection of Heavy Metals in Food

Presently, hazardous heavy metal ions such as Hg^{2+}, Pb^{2+}, Cd^{2+}, sulfite ions, sulfur ions, and nitrite ions have garnered huge attention due to their heavy risk to human and animal health (Ding et al., 2021). Contamination of food and water with these hazardous heavy metal ions will give rise to several disorders, such as reproductive malfunction, minamata, kidney breakdown, heart problems, and neurological disorders, even a trace of these hazardous metal ions in food and water can cause serious illness (Eddaif et al., 2019;

Dhumaleet al., 2020). Multiple detection techniques were utilized to detect such hazardous elements like atomic absorption, chromatography, electrical analysis method, fluorescence spectrophotometry, and colorimetric analyses.

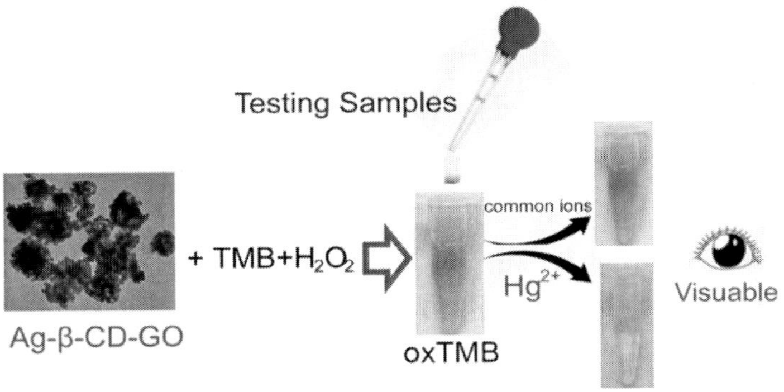

Reprinted with permission from Xing et al. (2021). Copyright © 2021, American Chemical Society.

Figure 8. Schematic diagram of Ag-β-CD-GO for detection of Hg^{2+}.

Xing et al., (2021) worked on a standardized redox-active self-assembly method to construct Ag-β-cyclodextrin-graphene oxide (GO) nanozyme (Ag-β-CDGO) for Hg^{2+} detection (Figure 8). The as-synthesized nanozyme displayed high stability, non-toxicity, and good surface area with strong biocompatibility and also exhibited excellent hydrogen-bonding capability of β-cyclodextrin with GO. The nanozyme exhibits excellent POD behavior that could accelerate TMB oxidation with the aid of H_2O_2. Moreover, it displays satisfied steady-state kinetic behavior to TMB and H_2O_2 with Michaelis-Menten constants (K_m) and maximum reaction velocity (V_{max}) of 3.3 $mmol{\cdot}L^{-1}$/2.45 × 10^{-8} $mol{\cdot}L^{-1}{\cdot}s^{-1}$ and 0.13 $mmol{\cdot}L^{-1}$ /2.52 × 10^{-8} $mol{\cdot}L^{-1}{\cdot}s^{-1}$ respectively. Hence, the above enzyme mimicking system could visually detect toxic Hg^{2+} through a color transformation from blue to transparent with varying Hg^{2+} concentrations. Moreover, the practical applicability of the nanozyme for Hg^{2+} detection was verified in real samples like water and fruit juice.

Adegoke et al., (2021) reported a colorimetric sensing application for nitrite based on the POD behavior of (AuNP)/cerium oxide (CeO_2) attached graphene oxide (GO). Selective and sensitive detection of nitrite ions was executed in the linear concentration range of 100 to 5000 μM with a limit of detection of 4.6 μM. Herein, a unique green color of the reaction solution was observed in the interaction of nitrite ions with the AuNP-CeO_2 NP@GO hybrid nanozyme.

Arsenic poisoning in drinking water is a serious concern to public health. Thus, it is of utmost importance to develop suitable techniques for the detection and removal of hazardous arsenate from the environment (Ma et al., 2014, Rizwan et al., 2014). Wen et al., (2019) worked on cobalt oxyhydroxide (CoOOH) nanoflakes exhibiting superb POD mimic activity for the detection of arsenate in water and food. The cobalt-based nanozyme effectively catalyzes the ABTS oxidation in presence of hydrogen peroxide to form a green oxidized solution. Herein, arsenate ion electrostatically binds to CoOOH nanoflakes thereby forming As-O bond interaction. Hence, the POD-like activity of CoOOH nanozyme declined on interaction with arsenate. Thus, a rapid detection method for arsenate was developed based on ABTS catalysis thereby obtaining a LOD value of 3.72 ppb. Moreover, the nanozyme also served a dual purpose for electrochemical detection of arsenate through conversion of the colorimetric sensor to an electrochemical sensor by utilization of CoOOH nanoflake-modified electrode. The detection limit obtained by the electrochemical sensor was found to be 56.1 ppt.

5.5. Detection of Antibiotics

Antibiotics as human and veterinary drugs have been extensively applied in the therapy of bacterial infection and the prevention of diseases of livestock (Jiang et al., 2019; Ahmed et al., 2020). A widely used antibiotic like kanamycin (KAN) has been commercially used in the treatment of infections in humans and animals. Kanamycin residue polluted the environment through contamination of the food chain, which causes serious side effects, such as earshot damage and nephrotoxicity. To date, kanamycin detection in food has drowned tremendous attention from researchers. Therefore, several techniques have been developed that include HPLC and ELISA for KAN detection (Nguyen et al., 2019; Bajkacz et al., 2020; Xiong et al., 2020). Despite the great accuracy of these technologies, they have a number of drawbacks, including the requirement for highly skilled employees, expensive costs, and time-consuming procedures. Antibiotic contamination residues may occur in numerous food products, including fruits, eggs, chicken, meat, milk, etc. (Han et al., 2019; Hendrickson et al., 2019). Considering the growing threat, the development of new technologies for sensing antibiotic drug in daily food products have been becoming a growing demand.

Tang et al., (2021) designed aptamer-enhanced WS_2 nanosheets as a colorimetric sensor based on its POD-mimicking activity for selective detection of KAN in food. The POD activity increases as a result of the modification of nanosheets through the aptamers which increases the binding affinity of TMB to the nanozyme. The LOD value obtained for colorimetric sensing of KAN was found to be 0.6 μM. Furthermore, the aptamer sensor displayed excellent selectivity against other contaminating antibiotics. Moreover, the residues of antibiotics exhibited a high probability of spreading in agricultural and livestock husbandry products. Hence, this sensing platform provided an effective method for KAN detection in honey, pork, and milk.

Electrochemical detection of KAN is also in recent trend among many researchers. Wang et al., (2016) successfully designed an electrochemical detection technique for KAN through the combination of aptamers and catalytically active Au NP nanozymes. The ssDNA when adsorbed to the Au NPs surface forms Au-S bonds, thereby blocking the peroxidase-mimicking activity of Au NPs. But, on conjugating the kanamycin with the aptamer, the active sites of Au NP become free to perform a redox reaction with H_2O_2, reducing thionine to ox-thionine. Through the differential pulse voltammetry analysis, the sensitive detection of kanamycin could be easily quantified by recording the electrochemical signal. Furthermore, this conventional technique was effectively applied to detect kanamycin in a real sample (honey).

Commercially available honey is highly contaminated with tetracyclines (TCs) antibiotics above a considerable limit permitted by WHO. Therefore, the use of these commercial products may lead to a threat to human health. Considering the threat, Sheng et al., (2020) worked on TC (tetracycline) detection based on aptamer-linked Au-NP in honey samples. Here, Au NPs act as POD mimic and catalyzes TMB oxidation with the assistance of H_2O_2 forming an oxidized TMB blue-colored solution. This technique was very much suitable to detect TC in honey by recording the change in color of oxidized TMB solution and absorbance at 652 nm.

He et al., (2020) worked on Au@SiO_2 nanozyme exhibiting excellent POD-like activity for rapid detection of sulfadiazine (SDZ) through a novel peroxidase-like nanozyme-linked immunosorbent assay. The detection limit (LOD) obtained in SDZ was 0.2 mg/L. Moreover, the detection was performed in beef samples spiked with SDZ where recoveries percentage in the 78.00% to 90.96% range was obtained.

5.6. Detection of H_2O_2 as a Food Contaminant

Hydrogen peroxide is known as an inorganic molecule that is broadly used in food production, medicinal purposes, pharmaceuticals, and paper manufacturing (Zhang et al., 2017). H_2O_2 has been misused in marine food products like jellyfish, shark fin, and shrimp for whitening, deodorizing and improving of appearance. H_2O_2 also have been used for the preservation of food and milk products to extend shelf life thereby keeping the raw milk fresh (Salih et al., 2017). However, H_2O_2 containing milk significantly loses its nutritional value, and milk consumers may suffer from gastrointestinal and neurodegenerative disorders (Dai et al., 2017). Consumption of food containing excessive H_2O_2 can irreversibly damage the central nervous system if the concentration exceeds 0.05% during food production declared by the United States Food and Drug Administration (FDA).

Zhang et al., (2018) designed Pd (palladium) nanoparticles supported mesoporous carbon (Pd NPs/meso-C) for TMB oxidation in presence of H_2O_2. Dispersion of Pd nanoparticles on mesoporous carbon increases the specific surface area of the nanocomposite thereby exhibiting excellent catalytic performance (K_{cat} value) higher than that of Pd NPs alone. The as-synthesized nanozyme could be applied as a chemosensor for H_2O_2 detection. Moreover, the nanozyme was also performed as a paper-based sensor by simply fixing a solution of TMB and Pd NPs/meso-C on an ordinary filter paper. The combination of sensor and test strips could be accessed on a smartphone and color-scanning app for quantitatively detecting H_2O_2. Moreover, the sensing platform demonstrates its potential for target detection for detecting H_2O_2 in milk products.

Liu et al., (2018) designed Fe/CuSn$(OH)_6$ microspheres behaving as POD mimic that catalyzes TMB oxidation with H_2O_2 assistance to form a blue oxidized solution. The reaction mechanism further confirms the formation of •OH radical, assigning to H_2O_2 decomposition. Hence, a detection strategy of H_2O_2 was designed based on the POD activity of Fe/CuSn$(OH)_6$ microspheres in a certain concentration (30 μM -1000 μM) with a limit of detection of 9.49 μM. This method qualitatively and quantitatively detects H_2O_2 in milk samples having recoveries in the range of 98.77-102.53%.

5.7. Detection of Food Allergy Proteins

Food allergy, originated from immunoglobulin E (IgE)/cell-mediated humoral immune type I hypersensitivity responses when got exposed to a class of life-menacing allergens present in food (He et al., 2018). According to the European Academy of Allergology and Clinical Immunology (EAACI), 5% of adults and 8% of children are globally affected by food allergens (Holzhauser 2018; Sicherer et al., 2018). Hence, the successful elimination the food allergens while processing food is of great challenge to the food industry and clinical allergology (Fu et al., 2019). Though some food might be rich in high-quality nutrients, the presence of food allergens might reduce the digestive capability of that food. Many food items like milk, peanuts, soybeans, eggs, sesame, wheat, tuna, salmon fish, shellfish and tree nuts commonly walnut, cashew, almond, pistachio and others contain a certain number of allergens reported by Food Allergen Research & Education (FARE) (https://www.foodallergy.org/allergens, 2021). Among all, milk is considered a common food that causes allergic reactions on consumption due to its milk proteins (Høst et al., 2020). Milk is rich in various whey proteins, numerous proteins, and four caseins that have been recognized as allergens. Considering these issues, various analytical methods such as polymerase chain reaction (PCR), LC−MS/MS, and other immunoassays have been designed for the successful determination of allergens in food (Chen et al., 2016; Su et al., 2019; Yang et al., 2014). Among all, enzyme-linked immunosorbent assay (ELISA) techniques are extensively applied for the rapid sensing of food allergens (Wu et al., 2019).

Zhang et al., (2021) developed (CeO_2) nanozyme as a peroxidase mimic for the detection of food allergy proteins based on surface-enhanced Raman scattering (SERS) immunoassay (Figure 9). For the detection of food allergens, α-LA allergen was designated as a detecting analyte. Here, the surface of a CeO_2 nanozymes was modified by polyclonal antibody (Ab2) to form CeO_2 nanozymes@Ab2 as the addition of Ab2 activated the nanozyme surface for specific recognition toward α-LA. CeO_2 nanozymes react with Raman inactive reporter (leucomalachite green (LMG)) thereby oxidizing it to malachite green (MG) with the assistance of H_2O_2. The Raman signal was further intensified through the interaction of CeO_2 with Raman "hot spot" Au NPs (gold nanoparticles). The Raman spectrometer captured the signals of MG for effective sensing of α-LA. Hence, the nanozyme-based SERS technique efficiently reduces the background interferences and also improves the detection capability via a secondary Raman signal amplification and the formed Raman "hot spot". Moreover, the conventional immunoassay method displayed several advantages like ease of operation, enhanced sensitivity and specificity, and extraordinary utility for sensing food allergens.

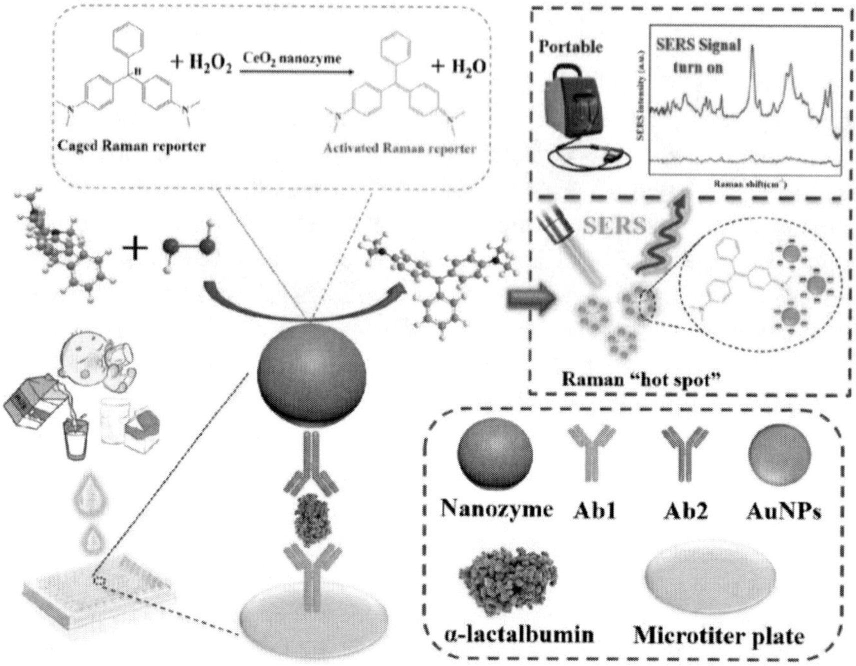

Reprinted with permission from Zhang et al. (2021). Copyright © 2021, American Chemical Society.

Figure 9. Schematic diagram of POD-like nanozyme for SERS immunoassay of food allergy proteins.

5.8. Detection of Harmful Chemicals and Additives in Food

Detection of harmful chemicals and additives in food is a huge challenge for researchers all over the world. Formaldehyde (HCHO) commonly known as a preservative is a harmful chemical that has a wide range of applications in many commercial treatments like resin manufacturing, synthetic plastic production, leather processing, etc. (Sayed et al., 2016; Deng et al., 2020). HCHO is applied in many chemical transformations like catalytic reduction to methanol, formic acid and oxidation to carbon dioxide. Formaldehyde exhibits carcinogenic and mutagenic behavior when binds to DNA, proteins, and other biomolecules thereby affecting their biological properties (Hauptmann et al., 2020). HCHO has a peculiar smell, and a concentration of more than 0.1 mg m^{-3} causes irritating eyes, throat aches, sickness, etc.

Zhao et al., (2021) developed a MnO_2 nanozyme with oxidase-mimicking activity in the existence of o-phenylenediamine (OPD) for fluorometric detection of formaldehyde. The OXD mimic MnO_2 in presence of a substrate OPD detects HCHO, resulting in the formation of a yellow solution containing 2,3-diaminophenazine (DAP). A bright yellow fluorescence signal at an emission wavelength of 560 nm was observed. HCHO readily reacts with o-phenylenediamine to form Schiff bases that reduce the oxidative coupling of OPD to DAP which was further confirmed by the quenching of fluorescence emission in the presence of HCHO. Here, the as-synthesized nanozyme selectively detects formaldehyde in a certain concentration range of 0.8-100 μM with a LOD value of 6.2×10^{-8} M (Figure 10). Moreover, the sensing probe can be further applied for HCHO detection in various samples like air, beer, and food items with good recovery performance (Figure 10).

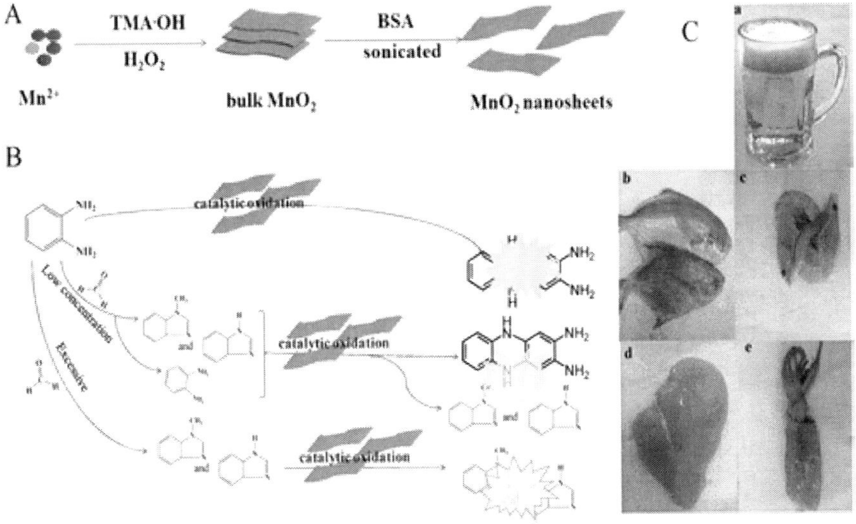

Reprinted with permission from Zhao et al. (2021). Copyright © 2021, American Chemical Society.

Figure 10. (A) Schematic diagram for MnO_2 nanosheet synthesis (B) Detection of HCHO based on OXD mimicking activity of MnO_2 nanosheets and DAP formation. (C) Photographs of real samples.

Melamine is widely used as an additive in different packaged food items by food processing units which creates a harmful effect on human health, hence melamine detection in food samples is of specific attention. Kumar et al., (2016) synthesized POD-mimicking silver nanoparticles (Ag NPs) for colorimetric detection of melamine in milk samples. Melamine present in milk combines with Ag NPs, leading to the aggregation of nanoparticles. The POD behavior of the Ag NPs declined as a result of its aggregation. The detection of melamine was monitored by recording the absorbance value through UV-vis spectrometry with a limit of detection of 0.04 mg L^{-1}.

Conclusion and Future Challenges

In comparison to natural enzymes, nanozymes exhibit several advantages:

(A) The catalytic properties of nanozymes are influenced by numerous factors such as surface modifications, functionalization, size, shape, and chemical composition. Hence, different techniques are developing consistently so as to modify the properties of nanozymes for increasing their catalytic activity.
(B) Nanozymes hold exclusive properties, such as magnetism, fluorescence, photothermal and stability which award nanozymes with outstanding controllability.
(C) The nanozymes with enzyme-mimicking activities possesses unique physicochemical or biological properties. Hence, the nanozyme could be tuned with various techniques to design superb nanomaterials with multifunctional behavior.

Moreover, the nanozymes mimicking natural enzymes have several advantages like upgraded stability and inexpensive synthesis procedure, with potential application in multiple fields. Even though there exist countless advances in this field, some problems still persist which are indefinable and need additional examination. Nanozymes although mimics natural enzymes suffer from several downfalls. The increment of substrate specificity of nanozymes still remains a challenge to several researchers. Moreover, nanozymes undergoes redox reaction in a catalytic cycle which further limits their application in biomedical applications.

To date, understanding the catalytic mechanisms at the atomic level is limited as it is very much important to decode them for the further achievement of high catalytic activity. Therefore, the catalytic mechanism at an atomic level is of utmost necessity to design new techniques for further improvement in the scientific field. Furthermore, the nanozymes with high catalytic efficiency in the certain reaction may not be so much efficient in bio-medical applications. In some cases, the nanozymes, when applied in ROS-mediated disease treatment, cause some probable side effects due to their multi-functional catalytic sites thereby hindering the effectiveness of the therapeutic treatment. Hence, nanozymes need to be cautiously designed.

Sooner or later, there may be some probability of utilizing additional appropriate standardized methods to identify the extent of enzyme-mimic activity of nanozymes. At the same time, some reference standard materials should be constructed to deliver valuable comparison standards for various nanozymes. As a whole, we expect this chapter will provide readers with a complete understanding of nanozymes. In addition to never-ending exploration, nanozymes or artificial enzymes simplifies their applications in every possible field including nanomedicine with an auspicious outlook to benefit human health.

References

Abhijith, K. S., Ragavan, K. V., Thakur, M. S. (2013). Gold nanoparticles enhanced chemiluminescence-a novel approach for sensitive determination of aflatoxin-B1. *Analytical Methods*, 4838-4845.

Adegoke, O., Zolotovskaya, S., Abdolvand, A., Daeid, N. N. (2021). Rapid and highly selective colorimetric detection of nitrite based on the catalytic-enhanced reaction of mimetic Au nanoparticle-CeO_2 nanoparticle-graphene oxide hybrid nanozyme. *Talanta*, 121875.

Ahmed, S., Ning, J., Peng, D., Chen, T., Ahmad, I., Ali, A., Lei, Z., Shabbir, M. A. B., Chenh, G., Yuan, Z. (2020). Current advances in immunoassays for the detection of antibiotics residues: A review. *Food and Agricultural Immunology*, 268-290.

Alshannaq, A., Yu, J.-H. (2017). Occurrence, Toxicity, and Analysis of Major Mycotoxins in Food. *International Journal of Environmental Research and Public Health*, 632.

Alsharif, A. M. A., Choo, Y.-M., Tan, G.-H. (2019). Detection of five mycotoxins in different food matrices in the Malaysian market by using validated liquid chromatography-electrospray ionization triple quadrupole mass spectrometry. *Toxins*, 196.

Altunbas, O., Ozdas, A., Yilmaz, M. D. (2020). Luminescent detection of Ochratoxin A using terbium chelated mesoporous silica nanoparticles. *Journal of Hazardous Materials*, 121049.

Andrade, P. D., Dantas, R. R., Moura-Alves, T. L. d. S. d., Caldas, E. D. (2017). Determination of multi-mycotoxins in cereals and of total fumonisins in maize products using isotope labelled internal standard and liquid chromatography/tandem mass spectrometry with positive ionization. *Journal of Chromatography A*, 138-147.

Aragay, G., Pino, F., Merkoci, A. (2012). Nanomaterials for sensing and destroying pesticides. *Chemical Rev*iews 5317-5338.

Asati, A., Santra, S., Kaittanis, C., Nath, S., Perez, J. M. (2009). Oxidase-like activity of polymer-coated cerium oxide nanoparticles. *Angewandte Chemie*, 2308-2312.

Bajkcz, S., Felis, E., Kycia-Slocka, E., Harnisz, M., Korzeniewska, E. (2020). Development of a New SLE-SPE-HPLC-MS/MS Method for the Determination of Selected Antibiotics and their Transformation Products in Anthropogenically Altered Solid Environmental Matrices. *Science of Total Environm*ent, 138071.

Baldim, V., Bedioui, F., Mignet, N., Margaill, I., and Berret, J. F. (2018). The enzyme-like catalytic activity of cerium oxide nanoparticles and its dependency on Ce3+ surface area concentration. *Nanoscale*, 6971-6980.

Benedetti, T. M., Andronescu, C., Cheong, S., Wilde, P., Wordsworth, J., Kientz, M., Tilley, R. D., Schuhmann, W., Gooding, J. J. (2018). Electrocatalytic nanoparticles that mimic the three-dimensional geometric architecture of enzymes: Nanozymes. *Journal of the American Chemical Society*, 13449–13455.

Cao, S., Tao, F. F., Tang, Y., Li, Y., and Yu, J. (2016). Size- and shape-dependent catalytic performances of oxidation and reduction reactions on nanocatalysts. *Chemical Society Reviews*, 4747-4765.

Chen, Q., Zhang, J., Ke, X., Lai, S., Li, D., Yang, J., Mo, W., Ren, Y. (2016). Simultaneous quantification of α-lactalbumin and β-casein in human milk using ultra-performance liquid chromatography with tandem mass spectrometry based on their signature peptides and winged isotope internal standards. *Biochimica et Biophysica Acta*, 1122-1127.

Cheng, N., Song, Y., Zeinhom, M. M. A., Chang, Y. C., Sheng, L., Li, H., Du, D., Li, L., Zhu, M. J., Luo, Y., Xu, W., Lin, Y. (2017). Nanozyme-mediated dual-immunoassay integrated with a smartphone for use in simultaneous detection of pathogens. *ACS Applied Materials & Interfaces*, 40671-40680.

Chung, K., Coyle, E. M., Jani, D., King, L. R., Bhardwaj, R., Fries, L., Smith, G., Glenn. G., Golding, H., Khurana, S. (2015). ISCOMATRIXTM adjuvant promotes epitope spreading and antibody affinity maturation of influenza A H7N9 virus-like particle vaccine that correlates with virus neutralization in humans. *Vaccine*, 3953-3962.

Dai, H. X., Lu, W. J., Zuo, X. W., Zhu, Q., Pan, C. J., Niu, X. Y., Liu, J. J., Chen, H. L., Chen, X. G. (2017). A novel biosensor based on boronic acid functionalized metal-organic frameworks for the determination of hydrogen peroxide released from living cells. *Biosensors Bioelectronics*, 131-137.

Deng, L., Liu, Q., Lei, C., Zhang, Y., Huang, Y., Nie, Z., Yao, S. (2020). Fluorometric and colorimetric dual-readout assay for histone demethylase activity based on formaldehyde inhibition of Ag^+-triggered oxidation of o-phenylenediamine. *Analytical Chem*istry, 9421-9428.

Derrick, J., Hollinghurst, P., O'Brien, S., Elviss, N., Allen, D. J., Gómara, M. (2021). Measuring transfer of human norovirus during sandwich production: Simulating the role of food, food handlers and the environment. *International Journal of Food Microbiology*, 109151.

Dhumale, V. A., Gangwar, R. K., Pande, N. (2020). Importance of gold nanoparticles for detection of toxic heavy metal ions and vital role in biomedical applications. *Materials Research Innovations*, 1-9.

Ding, R., Cheong, Y. H., Ahamed, A., Lisak, G. (2021). Heavy Metals Detection with Paper-Based Electrochemical Sensors. *Analytical Chemistry*, 1880–1888.

Dong, G. Y., Peng, C., Luo, J., Wang, C. M., Han, L., Wu, B., Ji, J., He. H., (2015). Adamantane-resistant influenza viruses in the world (1902-2013): Frequency and distribution of M2 gene mutations. *PloS One*, e0119115.

Dong, H., Fan, Y., Zhang, W., Gu, N., Zhang Y. (2019). Catalytic Mechanisms of Nanozymes and Their Applications in Biomedicine. *Bioconjugate Chemistry*, 1273-1296.

Dong, J., Song, L., Yin, J., He, W., Wu, Y., Gu, N., and Zhang, Y. (2014). Co3O4 nanoparticles with multi-enzyme activities and their application in the immunohistochemical assay. *ACS Applied Materials & Interfaces*, 1959-1970.

Eddaif, L., Shaban, A., Telegdi, J. (2019). Sensitive detection of heavy metal ions based on the calixarene derivatives-modified piezoelectric resonators: a review. *International Journal of Environmental Analytical Chemistry*, 824-853.

Edite Bezerra da Rocha, M., Freire, F. D. C. O., Erlan Feitosa Maia, F., Izabel Florindo Guedes, M., Rondina, D. (2014). Mycotoxins and their effects on human and animal health. *Food Control*, 159-165.

Egmond, H. P. V, Schothorst, R. C., Jonker, M. A. (2007). Regulations relating to mycotoxins in food: Perspectives in a global and European context. *Analytical and Bioanalytical Chemistry*, 147-157.

Estes, M. K., Ettayebi, K., Tenge, V. R., Murakami, K., Karandikar, U., Lin, S. C., Ayyar, B. V., Cortes-Penfield, N. W., Haga, K., Neill, F. H., Opekun, A. R. (2019). Human norovirus cultivation in non-transformed stem cell-derived human intestinal enteroid cultures: success and challenges. *Viruses*, 638.

Fan, J., Yin, J., Ning, B., Wu, X., Hu, Y., Ferrari, M., Anderson, G., Wei, J., Zhao, Y., Nie, G. (2011). Direct evidence for catalase and peroxidase activities of ferritin-platinum nanoparticles. *Biomaterials*, 1611-1618.

Fan, K., Wang, H., Xi, J., Liu, Q., Meng, X., Duan, D., Gao, L., Yan, X. (2017). Optimization of Fe_3O_4 nanozyme activity via single amino acid modification mimicking an enzyme active site. *Chemical Communications*, 424-427.

Fan, L., Wu, H., Lou, D., Zhang, X., Zhu, Y., Gu, N., Zhang, Y. (2018). A novel AuNPs-based glucose oxidase mimic with enhanced activity and selectivity constructed by molecular imprinting and O_2-containing nanoemulsion embedding. *Advanced Materials Interfaces*, 1801070.

Farka, Z., Čunderlova, V., Horaćkova, V., Pastucha, M., Mikusova, Z., Hlavaćek, A., Skladal, P. (2018). Prussian blue nanoparticles as a catalytic label in a sandwich nanozyme-linked immunosorbent assay. *Analytical Chemistry*, 2348-2354.

Food Allergy Research & Education (FARE). *Common Allergens;* FARE: McLean, VA, 2021; https://www.foodallergy.org/allergens.

Fu, H. Y., Li, H. D., Wang, B., Cai, J. L., Guo, J.W., Cui, H. P., Zhang, X. B., Yu, Y. J. (2016). Quantification of acid metabolites in complex plant samples by using second-order calibration coupled with GC-mass spectrometry detection to resolve the influence of seriously overlapped chromatographic peaks. *Analytical Methods*, 747-755.

Fu, H., Hu, O., Xu, L., Fan, Y., Shi, Q., Guo, X., Lan, W., Yang, T., Xie, S., She, Y. (2019). Simultaneous Recognition of Species, Quality Grades, and Multivariate Calibration of Antioxidant Activities for 12 Famous Green Teas Using Mid- and Near-Infrared Spectroscopy Coupled with Chemometrics. *Journal of Analytical Methods in Chemistry*, 4372395.

Fu, L., Cherayil, B. J., Shi, H., Wang, Y., Zhu, Y. (2019). Detection and quantification methods for food allergens. *Food Allergy*, 69–91.

Gao, L., Zhuang, J., Nie, L., Zhang, J., Zhang, Y., Gu, N., Wang, T., Feng, J., Yang, D., Perrett, S. (2007). Intrinsic peroxidase-like activity of ferromagnetic nanoparticles. *Nature Nanotechnology*, 577–583.

Gao, Y., Zhou, Y. Z., Chandrawati, R. (2020). Metal and metal oxide nanoparticles to enhance the performance of enzyme-linked immunosorbent assay (ELISA). *ACS Applied Nano Materials*, 1–21.

Ghosh, S., Roy, P., Karmodak, N., Jemmis, E. D., and Mugesh, G. (2018) Nanoisozymes: crystal-facet-dependent enzyme-mimetic activity of V_2O_5 nanomaterials. *Angewandte Chemie*, 4510−4515.

Han, M., Cong, L., Wang, J., Zhang, P., Jin, Y., Zhao, R., Yang. C., He. L., Feng, X., Chen, Y. (2019). An octuplex lateral flow immunoassay for rapid detection of antibiotic residues, aflatoxin M1 and melamine in milk. *Sensors and Actuators B: Chemical*, 94-104.

Hauptmann, M., Stewart, P. A., Lubin, J. H., Freeman, L. E. B., Hornung, R. W., Herrick, R. F., Hoover, R. N., Fraumeni Jr, J. F., Blair, Aaron., Hayes, R. B. (2020). Mortality from lymphohematopoietic malignancies and brain cancer among embalmers exposed to formaldehyde. *Journal of the National Cancer Institute*, 1518-1519.

He, J., Liu, G., Jiang, M., Xu, L., Kong, Feifan., Xu, Z. (2020). Development of novel biomimetic enzyme-linked immunosorbent assay method based on Au@SiO_2 nanozyme labelling for the detection of sulfadiazine. *Food and Agricultural Immunology*, 341-351.

He, S., Li, X., Wu, Y., Wu, S., Wu, Z., Yang, A., Chen, H. (2018). Highly sensitive detection of bovine β-lactoglobulin with wide linear dynamic range based on platinum nanoparticles probe. *Journal of Agricultural and Food Chemistry*, 11830-11838.

He, W. W., Zhou, Y. T., Warner, W. G., Hu, X., Zheng, Z., Boudreau, M. D., Yin, J. J. (2013). Intrinsic catalytic activity of Au nanoparticles with respect to hydrogen peroxide decomposition and superoxide scavenging. *Biomaterials*, 765–773.

Heckert, E. G., Karakoti, A. S., Seal, S., Self, W. T. (2008). The role of cerium redox state in the SOD mimetic activity of nanoceria. *Biomaterials*, 2705−2709.

Hendrickson, O. D., Zvereva, E. A., Shanin, I. A., Zherdev, A. V., Dzantiev, B. B. (2019), Development of a multicomponent immunochromatographic test system for the detection of fluoroquinolone and amphenicol antibiotics in dairy products. *Journal of the Science of Food and Agriculture*, 3834-3842.

Hizir, M. S., Top, M., Balcioglu, M., Rana, M., Robertson, N. M., Shen, F., Sheng, J., and Yigit, M. V. (2016). Multiplexed activity of perAuxidase: DNA-capped AuNPs act as adjustable peroxidase. *Analytical Chemistry*, 600-605.

Holzhauser, T. (2018). Protein or No Protein? Opportunities for DNA Based Detection of Allergenic Foods. *Journal of Agricultural and Food Chemistry*, 9889-9894.

Nguyen, T. A. H., Pham, T. N. M., Le, T. B., Le, D. C., Tran, T. T. P., Nguyen, T. Q. H., Nguyen, T. K. T., Hauser, P. C., Mai, T. D. (2019). Cost-effective Capillary Electrophoresis with Contactless Conductivity Detection for Quality Control of Beta-lactam Antibiotics. *Journal of Chromatography A*, 360356.

Ostry, V., Malir, F., Toman, J., Grosse, Y. (2017). Mycotoxins as human carcinogens-the IARC Monographs classification. *Mycotoxin Research*, 65-73.

Pagkali, V., Petrou, P. S., Makarona, E., Peters, J., Haasnoot, W., Jobst, G., Moser, I., Gajos, K., Budkowski, A., Economou, A., Misiakos, K., Raptis, I., Kakabakos, S. E. (2018). Simultaneous determination of aflatoxin B_1, fumonisin B_1 and deoxynivalenol in beer samples with a label-free monolithically integrated optoelectronic biosensor. *Journal of Hazardous Materials*, 445-453.

Palizban, A. A., Sadeghi-Aliabadi, H., Abdollahpour, F. (2010), Effect of cerium lanthanide on Hela and MCF-7 cancer cell growth in the presence of transferrin. *Research in Pharmaceutical Sciences*, 119–125.

Pereira, C. G., Andrade, J., Ranquine, T., Moura, I. N. de., Rocha, R. A. da., Furtado, M. A. M., Bell, M. J. V., Anjos, V. (2018). Characterization and detection of adulterated whey protein supplements using stationery and time-resolved fluorescence spectroscopy. *Lebensmittel-Wissenschaft & Technologie*, 180-186.

Pirmohamed, T., Dowding, J. M., Singh, S., Wasserman, B., Heckert, E., Karakoti, A. S., King, J. E. S., Seal, S., Self, W. T. (2010). Nanoceria exhibit redox state-dependent catalase mimetic activity. *Chemical Communications*, 2736-2738.

Pope, C. N. (1992). Organophosphorus pesticides: Do they all have the same mechanism of toxicity? *Journal of Toxicology and Environmental Health, Part B*, 161-181.

Rizwan, S., Naqshbandi, A., Farooqui, Z., Khan, A. A., Khan, F. (2014). Protective effect of dietary flaxseed oil on arsenic-induced nephrotoxicity and oxidative damage in rat kidney. *Food and Chemical Toxicology*, 99-107.

Salih, M. A. N., Yang, S. M. (2017). Common milk adulteration in developing countries cases study in China and Sudan: a review. *Advances in Dairy Research*, 192.

Sayed, S. EI, Pascual, L., Licchelli, M., Martínez-Máñez, R., Gil, S.; Costero, A. M., Sancenón, F. (2016). Chromogenic detection of aqueous formaldehyde using functionalized silica nanoparticles. *ACS Applied Materials & Interface*, 14318-14322.

Shen, X., Liu, W., Gao, X., Lu, Z., Wu, X., and Gao, X. (2015). Mechanisms of oxidase and superoxide dismutation-like activities of gold, silver, platinum, and palladium, and their alloys: a general way to the activation of molecular oxygen. *Journal of the American Chemical Society*, 15882-15891.

Sheng, Y. M., Liang, J., Xie, J. (2020). Indirect Competitive Determination of Tetracycline Residue in Honey Using an Ultrasensitive Gold-Nanoparticle-Linked Aptamer Assay. *Molecules*, 2144.

Sicherer, S. H., Sampson, H. A. (2018). Food allergy: A review and update on epidemiology, pathogenesis, diagnosis, prevention, and management. *The Journal of Allergy and Clinical Immunology*, 41-58.

Singh, J., Mehta, A. (2020). Rapid and sensitive detection of mycotoxins by advanced and emerging analytical methods: A review. *Food Science & Nutrition*, 2183-2204.

Soon, J. M., Brazier, A. K. M., Wallace, C. A. (2020). Determining common contributory factors in food safety incidents- A review of global outbreaks and recalls 2008-2018. *Trends in Food Science and Technology*, 76-87.

Su, Y., Wu, D., Chen, J., Chen, G., Hu, N., Wang, H., Wang, P., Han, H., Li, G., Wu, Y. (2019). Ratiometric Surface Enhanced Raman Scattering Immunosorbent Assay of Allergenic Proteins via Covalent Organic Framework Composite Material Based Nanozyme Tag Triggered Raman Signal "Turn-on" and Amplification. *Analytical Chemistry*, 11687-11695.

Sun, Y., Zhao, C., Gao, N., Ren, J., and Qu, X. (2017). Stereoselective nanozyme based on ceria nanoparticles engineered with amino acids. *Chemistry-A European Journal*, 18146-18150.

Tang, Y., Hu, Y., Zhou, P., Wang, C., Tao, Han., Wu, Y. (2021). Colorimetric Detection of Kanamycin Residue in Foods Based on the Aptamer-Enhanced Peroxidase-Mimicking Activity of Layered WS_2 Nanosheets. *Journal of Agricultural and Food Chemistry*, 2884−2893.

Tarokh, A., Pebdeni, A. B., Othman, H. O., Salehnia, F., Hosseini, M. (2021). Sensitive colorimetric aptasensor based on g-C_3N_4@Cu_2O composites for detection of Salmonella typhimurium in food and water. *Microchimica Acta*, 87.

Todd, E. C., Greig, J. D., Bartleson, C. A., Michaels, B. S. (2007). Outbreaks where food workers have been implicated in the spread of foodborne disease. Part 3. Factors contributing to outbreaks and description of outbreak categories. *Journal of Food Protection*, 2199-221.

Turner, N. W., Subrahmanyam, S., Piletsky, S. A. (2009). Analytical methods for determination of mycotoxins: A review. *Analytica Chimica Acta*, 168-180.

Vallabani, N. V. S., Karakoti, A. S., Singh, S. (2017). ATP mediated intrinsic peroxidase-like activity of Fe_3O_4-based nanozyme: one-step detection of blood glucose at physiological pH. *Colloids and Surfaces, B*, 52-60.

Wang, C. H., Gao, J., Cao, Y. L., Tan, H. L. (2018). Colorimetric logic gate for alkaline phosphatase based on copper (II)-based metal-organic frameworks with peroxidase-like activity. *Analytica Chimica Acta*, 74-81.

Wang, C., Liu, C., Luo, J., Tian, Y., Zhou, N. (2016). Direct electrochemical detection of kanamycin based on peroxidase-like activity of gold nanoparticles. *Analytica Chimica Acta*, 75-82.

Wang, Q. Q., Wei, H., Zhang, Z. Q., Wang, E. K., Dong, S. J. (2018). Nanozyme: An emerging alternative to natural enzyme for biosensing and immunoassay. *Trends in Analytical Chemistry*, 218-224.

Wang, X. Y., Hu, Y. H. Wei, H. (2016). Nanozymes in bionanotechnology: From sensing to therapeutics and beyond. *Inorganic Chemistry Frontiers*, 41–60.

Weerathunge, P., Ramanathan, R., Torok, V. A., Hodgson, K., Xu, Y., Goodacre, R., Behera, B. K., Bansal, Vipul. (2019). Ultrasensitive Colorimetric Detection of Murine Norovirus Using NanoZyme Aptasensor. *Analytical Chem*istry, 3270-3276.

Wei, J. C., Yang, L. L., Luo, M., Wang, Y. T., Li, P. (2019). Nanozyme-assisted technique for dual mode detection of organophosphorus pesticides. *Ecotoxicology and Environmental Safety*, 17.

Wei, X. F., Zhou, W., Sanjay, S. T., Zhang, J., Jin, Q. J., Xu, F., Dominguez, D. C., Li, X. J. (2018). Multiplexed instrument-free bar-chart spin chip integrated with nanoparticle-mediated magnetic aptasensor for visual quantitative detection of multiple pathogens. *Analytical Chemistry*, 9888-9896.

Wen, S. H., Zhong, X. L., Wu, Y. D., Liang, R. P., Zhang, L., Qiu, J. D. (2019). Colorimetric Assay Conversion to Highly Sensitive Electrochemical Assay for Bimodal Detection of Arsenate Based on Cobalt Oxyhydroxide Nanozyme via Arsenate Absorption. *Analytical Chemistry*, 6487-6497.

Wong, E. L. S., Vuong, K. Q., Chow, E. (2021). Nanozymes for Environmental Pollutant Monitoring and Remediation. *Sensors*, 408.

Wu, J., Yang, Q., Li, Q., Li, H., Li, F. (2021). Two-Dimensional MnO_2 Nanozyme-Mediated Homogeneous Electrochemical Detection of Organophosphate Pesticides without the Interference of H2O2 and Colour. *Analytical Chemistry*, 4084-4091.

Wu, L., Li, G., Xu, X., Zhu, L., Huang, R., Chen, X. (2019). Application of nano-ELISA in food analysis: Recent advances and challenges. *TrAC Trends in Analytical Chemistry*, 140-156.

Wu, Y.N., Liu, P., Chen, J.S. (2018). Food safety risk assessment in China: Past, present and future. *Food Control*, 212-221.

Wua, S., Guoa, D., Xua, X., Panb, J., Niub, X. (2020). Colorimetric quantification and discrimination of phenolic pollutants based on peroxidase-like Fe_3O_4 nanoparticles. *Sensors and Actuators B: Chemical*,127225.

Xie, Y. L., Ning, M. G., Ban, J., Li, Q. (2017). Novel enzyme-linked aptamer assay for the determination of aflatoxin B_1 in peanuts. *Analytical Letters*, 2961-2973.

Xing, L., Zheng, X., Tang, Y., Zhou, X., Hao, J., Hu, L., Shen, J., Yan, Z. (2021). Ag-β-Cyclodextrin-Graphene Oxide Ternary Nanostructures with Peroxidase-Mimicking Activity for Hg^{2+} Detection. *ACS Applied Nano Materials*, 13807-13817.

Xiong, Y., Leng, Y. K., Li, X. M., Huang, X. L., Xiong, Y. H. (2020). Emerging Strategies to Enhance the Sensitivity of Competitive ELISA for Detection of Chemical Contaminants in Food Samples. *TrAC Trends in Analytical Chemistry*, 115861.

Xu, Z., Long, L.-l., Chen, Y.-q., Chen, M.-L., Cheng, Y.-H. (2021). A nanozyme-linked immunosorbent assay based on metal-organic frameworks (MOFs) for sensitive detection of aflatoxin B_1. *Food Chemistry*, 128039.

Xue, L., Jin, N., Guo, R., Wang, S., Qi, W., Liu, Y., Li, Y., Lin, J. (2021). Microfluidic Colorimetric Biosensors Based on MnO_2 Nanozymes and Convergence-Divergence Spiral Micromixers for Rapid and Sensitive Detection of Salmonella, *ACS Sensors*, 2883-2892.

Yang, A., Zheng, Y., Long, C., Chen, H., Liu, B., Li, X., Yuan, J., Cheng, F. (2014). Fluorescent immunosorbent assay for the detection of alpha-lactalbumin in dairy products with monoclonal antibody bio-conjugated with CdSe/ZnS quantum dots. *Food Chemistry*, 73-79.

Yang, H., Xiao, J., Su, L., Feng, T., Lvc, Q., Zhang, X. (2017). The oxidase-mimicking activity of the nitrogen-doped Fe3C@C composites. *Chemical Communications*, 3882-3885.

Yang, Y., Li, G., Wu, D., Liu, J., Li, X., Luo, P., Hu, N., Wang, H., Wu, Y. (2020). Recent advances on toxicity and determination methods of mycotoxins in foodstuffs. *Trends in Food Science and Technology*, 233–252.

Yoshioka, N., Ichihashi, K. (2008). Determination of 40 synthetic food Colors in drinks and candies by high-performance liquid chromatography using a short column with photodiode array detection. *Talanta*, 1408-1413.

Yu, L., Li, N. (2019). Noble Metal Nanoparticles-Based Colorimetric Biosensor for Visual Quantification: A Mini Review. *Chemosensors*, 53.

Yu, L., Ma, F., Zhang, L., Li, P. (2019). Determination of aflatoxin B_1 and B_2 in vegetable oils using Fe_3O_4/rGO magnetic solid phase extraction coupled with high-performance liquid chromatography fluorescence with post-column photochemical derivatization. *Toxins*, 621.

Zhang, R. Z., Chen, W. (2017). Recent advances in graphene-based nanomaterials for fabricating electrochemical hydrogen peroxide sensors. *Biosensors Bioelectronics*, 249-268.

Zhang, S. X., Xue, S. F., Deng, J., Zhang, M., Shi, G., Zhou, T. (2016). Polyacrylic acid-coated cerium oxide nanoparticles: An oxidase mimic applied for colorimetric assay to organophosphorus pesticides. *Biosensors & Bioelectronics*, 457-463.

Zhang, W., Hu, S., Yin, J., He, W., Lu, W., Ma, M., Gu, N., Zhang, Y. (2016). Prussian blue nanoparticles as multienzyme mimetics and reactive oxygen species scavengers. *Journal of the American Chemical Society*, 5860-5865.

Zhang, W., Niu, X., Li, Xin., He, Y., Song, H., Peng, Y., Pan, J., Qiu, F., Zhao, H., Lan, M. (2018). A smartphone-integrated ready-to-use paper-based sensor with mesoporous carbon-dispersed Pd nanoparticles as a highly active peroxidase mimic for H_2O_2 detection. *Sensors and Actuators B: Chemical*, 412-420.

Zhang, X., Li, G., Liu, J. Su, Z. (2021). Bio-inspired Nanoenzyme Synthesis and Its Application in A Portable Immunoassay for Food Allergy Proteins. *Journal of Agricultural and Food Chemistry*, 14751-14760.

Zhang, X., Wu, D., Zhou, X., Yu, Y., Liu, J., Hu, N., Wang, H., Li, G., Wu, Y. (2019). Recent progress in the construction of nanozyme-based biosensors and their applications to food safety assay. *Trends in Analytical Chemistry*, 115668.

Zhang, Y., Pei, F., Fang, Y., Li, P., Zhao, Y., Shen, F., Zou, Y., Hu, Q. (2019). Comparison of concentration and health risks of 9 Fusarium mycotoxins in commercial whole wheat flour and refined wheat flour by multi-IAC-HPLC. *Food Chemistry*, 763-769.

Zhang, Z., Liu, B., and Liu, J. (2017). Molecular imprinting for substrate selectivity and enhanced activity of enzyme mimics. *Small*, 1602730.

Zhang, Z., Su, B., Xu, H., He, Z., Zhou, Y., Chen, Q., Sun, Z., Cao, H., Liu, X. (2021). Enzyme cascade-amplified immunoassay based on the nanobody-alkaline phosphatase fusion and MnO_2 nanosheets for the detection of ochratoxin A in coffee. *RSC Advances*, 21760.

Zhang, Z., Zhang, X., Liu, B., and Liu, J. (2017). Molecular imprinting on inorganic nanozymes for hundred-fold enzyme specificity. *Journal of the American Chemical Society,* 5412-5419.

Zhao, Q., Shen, T., Liu, Y., Hu, X., Zhao, W., Ma, Z., Li, P., Zhu, X., Zhang, Y., Liu, M., Yao, S. (2021). Universal Nanoplatform for Formaldehyde Detection Based on the

Oxidase-Mimicking Activity of MnO_2 Nanosheets and the In Situ Catalysis-Produced Fluorescence Species. *Journal of Agricultural and Food Chemistry*, 7303-7312.

Zhou, P., Hu, O., Fu, H., Ouyang, L., Gong, X., Meng, P., Wang, Z., Dai, M., Guo, X., Wang, Y. (2019). UPLC-Q-TOF/MS-based untargeted metabolomics coupled with chemometrics approach for Tieguanyin tea with seasonal and year variations. *Food Chemistry*, 73-82.

Zhou, S., Xu, L., Kuang, H., Xiao, J., Xu, C. (2020). Immunoassays for rapid mycotoxin detection: State of the art. *Analyst*, 7088-7102.

Zhu, Y. Y., Wu, J. J. X., Han, L. J., Wang, X. Y., Li, W., Guo, H. C., Wei, H. (2020). Nanozyme sensor arrays based on heteroatom-doped graphene for detecting pesticides. *Analytical Chemistry*, 7444-7452.

Chapter 2

Nanomaterials-Based Artificial Enzymes as Sensors for Detecting Pesticides

Ashutosh Thakur[1,2,*] and Manash R. Das[1,2,†]

[1] Advanced Materials Group, Materials Sciences and Technology Division, CSIR-North East Institute of Science and Technology, Jorhat, Assam, India

[2] Academy of Scientific and Innovative Research (AcSIR), Ghaziabad, India

Abstract

In recent years, various nanomaterials have been extensively explored to convert substrates to products following the catalytic pathway of natural enzymes. These nanomaterial-based artificial enzymes, called nanozymes, can overcome several intrinsic drawbacks of natural enzymes owing to their high stability and low production cost. Nanozymes have been widely applied for diverse applications such as bioimaging, biosensing, immunoassays, therapeutics, and environmental toxicology.

The focus of this chapter is on the nanozymes-based detection of the pesticides. Pesticides are an important class of materials to protect crops from pests and harmful diseases. While pesticides are necessary to regulate plant growth, their uncontrolled use has become a serious threat to the environment and public health. Therefore, it is extremely important to develop simple, easily available, low-costs, and highly sensitive sensors to detect pesticides for safety and awareness. Research on various nanozymes has been found effective as colorimetric, fluorometric, and chemiluminescent sensors to detect different pesticides.

---

* Corresponding Author's Email: ashutoshthakur@neist.res.in.

† Corresponding Author's Email: mrdas@neist.res.in.

In: Emerging Environmental Applications of Nanozymes
Editors: Seema Nara and Smriti Singh
ISBN: 979-8-88697-552-9

This chapter deals with a brief overview of the principle of nanozymes, methods of preparation of different types of nanozymes, and their sensor applications for pesticides detection with a special focus on two-dimensional materials-based nanozymes.

Keywords: nanomaterials, nanozymes, sensors, pesticides, two-dimensional nanomaterials

1. Introduction

Pesticides are an important class of (moderately to highly) toxic chemical compounds extensively used to protect crops from weeds, fungus, insects, rodents, and harmful diseases. Pesticides are indispensable chemicals for modern agricultural practices as they boost plant growth, increase crop yields, and reduce post-harvesting wastes [1]. However, a continuous rise in food demands and the lack of sufficient public awareness of the toxic effects of pesticides have resulted in their uncontrolled use in the agriculture sector [2]. The mismanagement of pesticides has become a serious global issue, with 90% of the used pesticides being released as environmental pollutants [2, 3]. Some pesticides are hardly degraded, and as a result, the pesticide residues in soil eventually enter the food chain [4]. Prolonged consumption of pesticide-containing foods could cause serious health problems such as diabetes, asthma, leukaemia, etc. [5]. Since the worldwide consumption of pesticides is as high as two million tonnes per year [6], it is extremely important to develop efficient and affordable analytical methods for detecting the presence of pesticides in food, water, and soil.

Gas chromatography–mass spectrometry (GC-MS), high-performance liquid chromatography–mass spectrometry (HPLC-MS), capillary electrophoresis, and surface-enhanced Raman spectroscopy (SERS) are the most commonly employed analytical methods towards pesticides detection [7-10]. Although these methods offer the advantages of high sensitivity and specificity, the requirements of sophisticated instrumentation, complex operational process, and highly skilled human resources significantly restrict their applicability for rapid on-site detection of pesticides. Other techniques, for example, enzyme-linked immunosorbent assays [11, 12] and enzyme-based biosensors [13-15], have been widely explored for rapidly detecting pesticides. Nevertheless, natural enzymes and antibodies employed in these techniques are costly and become unstable at harsh operational conditions, limiting their scope in practical applications. Therefore, developing cost-effective sensors for the rapid detection of various pesticides and their biomarkers under a wide range of conditions is an important target.

Various nanomaterials have been used for colorimetric, photoluminescent or fluorescent-based sensing of pesticides. For instance, Sun et al. [16] developed a label-free simple colorimetric sensing technique for organophosphate (OP) pesticides in solutions and achieved a pM level limit of detection (LOD) based on the principle of change in surface plasmon color of aggregated gold nanoparticles (AuNPs). In this method, the catalytic hydrolysis of a nerve agent acetylthiocholine to cationic thiocholine by acetylcholine esterase (AChE) induces the agglomeration of lipoic acid capped AuNPs, resulting in a visible change in colour from red to steel-blue. In the presence of OPs as AChE inhibitors, the change in colour of an AuNPs-containing solution is diminished to different extents according to the used concentrations of OPs.

In a similar study, Liu et al. [17] developed a colorimetric and fluorescent dual-readout sensor based on rhodamine B capped AuNPs for detecting carbamate and OP pesticides in solutions. The mechanism of detection is based on the hydrolysis reaction of acetyl thiocholine to cationic thiocholine by AChE, causing the color change of rhodamine B capped AuNPs-containing solutions from red to blue due to NPs aggregation, accompanied by the strong fluorescence from rhodamine B molecules due to their detachment from Au surfaces. The presence of a variety of pesticides, including carbaryl, diazinon, malathion, and phorate, inhibit the activity of AChE, keeping the color of the AuNPs-containing solutions red and the fluorescence of rhodamine B molecules quenched due to their interactions with the NPs. Yi et al. [18] developed a label-free ultrasensitive photoluminescence (PL) sensor based on silicon quantum dots (SiQDs) for detection of OP pesticides. This sensing method includes two steps enzymatic reactions to transform acetylcholine chloride to choline by AChE and then oxidation of choline to produce H_2O_2 and betaine. The formed H_2O_2 can effectively quench the PL generated from SiQDs. When a pesticide is added, the enzymatic activity of AChE gets inhibited, resulting in a reduction in the formation of H_2O_2 with an enhancement of the PL from SiQDs. Thus, by quantifying the increase in the PL, the concentration of parathion, diazinon, carbaryl, and phorate pesticides can be measured with high precession.

Another class of nanomaterials with enzyme-like catalytic properties has been found promising as sensors for pesticides detection [19-21]. These artificial enzymes (nanozymes) possess the benefits of low cost, tolerance to harsh environmental conditions, large-scale production, and superior stability compared to natural enzymes [22]. Nanozymes employed for pesticides detection can be categorized into oxidoreductase and hydrolase mimics. Frequently used oxidoreductases in pesticides detection are peroxidase and oxidase mimics, which mainly work by catalytically amplifying sensor signals of enzymatic substrates [19]. Peroxidase mimic nanozymes catalyze H_2O_2 decomposition to hydroxyl (•OH) radicals and promote the oxidation of substrates such as 3,3′-diaminobenzidine (DAB), o-phenylenediamine (OPD), 3,3′,5,5′-tetramethylbenzidine (TMB), and 2,2-azobis(3-ethylbenzothiazolin-6-sulfonic acid) (ABTS) to produce fluorescent or/and coloured products. Oxidase mimics can *in-situ* generate H_2O_2 by catalysing the oxidation of substrates which can be used to detect analytes. In contrast, phosphatase mimic nanozymes are used as a class of hydrolases in pesticide detection assays, where they catalyse the hydrolysis of various OP pesticides.

In this chapter, we have provided a general overview of different methods of detection of pesticides using nanozymes as sensor materials. To better comprehend these methodologies by a wide spectrum of readerships, we also briefly summarize the types of pesticides and their toxicology, and nanomaterials as nanozymes in two different sections in the beginning.

2. Types of Pesticides and Their Toxicology

Pesticides can be broadly grouped into different chemical families. The main insecticides can be grouped into three families of chemicals: organochlorines, carbamates, and organophosphates [23]. Organochlorine (OC) insecticides, used worldwide due to their low cost, are a classof synthetic chlorinated compounds (e.g., DDT, DDD, chlorobenzilate, dicofol, dieldrin, eldrin, etc.) with severe environmental persistence. OC insecticides function by stimulating the central nervous system, inhibiting the Ca- and Mg-ATPase and calcium ion influx, disrupting the sodium/potassium balance of the nerve fiber, causing the release of neurotransmitters and affecting neurological outcomes of an individual [23].

Due to their bioaccumulation potential, many countries have banned OC insecticides. Organophosphate (OP) and carbamate (CM) insecticides have largely replaced OC insecticides. OP and CM compounds operate by inhibiting the activity of the acetylcholinesterase (AChE) enzyme [24, 25]. AChE catalyzes the breakdown of acetylcholine, which functions as neurotransmitters. Inhibiting the AChE enzyme by insecticides allows acetylcholine to transport nerve impulses indefinitely, causing severe weakness and paralysis symptoms. Important families of herbicides are triazines (e.g., simazine, atrazine), ureas (e.g., diuron), phenoxy (e.g., dichloroprop, mecoprop,), chloroacetanilide, and benzoic acid (e.g., dicamba, dichlorobenil) herbicides [26]. Triazine herbicides are known as insect chemosterilants. Higher concentrations of triazine herbicides interfere with the plant's photosynthesis process. Phenoxy compounds are mainly employed to kill agricultural weeds, and these compounds can be degraded by various microorganisms. The phenoxy and benzoic acid compounds work as growth hormones in plants and thus suppress the plant's nutrient transport system.

3. Enzyme-Mimetic Nanomaterials

It has been reported that several nanomaterials possess catalytic features similar to natural enzymes like oxidase, peroxidase, catalase, hydrolase, superoxide dismutase, etc. [27-29]. These functional nanomaterials having enzyme-like features are called nanozymes. Over the past two decades, numerous nanozymes such as metal oxide NPs, metal NPs, alloy NPs, carbon nanomaterials, and nanostructured composite materials have been synthesized and employed as nanozymes. These nanozymes equip the benefits of both nanomaterials and natural enzymes, thus being applicable for bioimaging, biosensing, immunoassays, therapeutics, environmental toxicology, etc. Among various nanomaterials, special focus has been given towards the utilization of two-dimensional (2D) nanomaterials, namely, hexagonal boron nitride (h-BN), graphene, graphene oxide (GO), reduced grpahene oxide (rGO), transition metal dichalcogenide and oxide (TMD/TMO), MXenes, metal nanosheets, and 2D metal-organic frameworks (MOFs) as nanozymes due to their unique chemical, physical, and biological properties [27-29]. More importantly, these 2D nanomaterials are also powerful for fabricating various 2D material-based functional nanocomposites towards nanozymatic applications. Figure 1 shows examples of various nanomaterials that have been explored as artificial enzymes.

Until now, most of the nanozymes used as sensors for pesticide detection are peroxidase and oxidase mimics. Therefore, a brief overview of the principle and functioning of peroxidase and oxidase mimic nanozymes using selected examples are described in the next section.

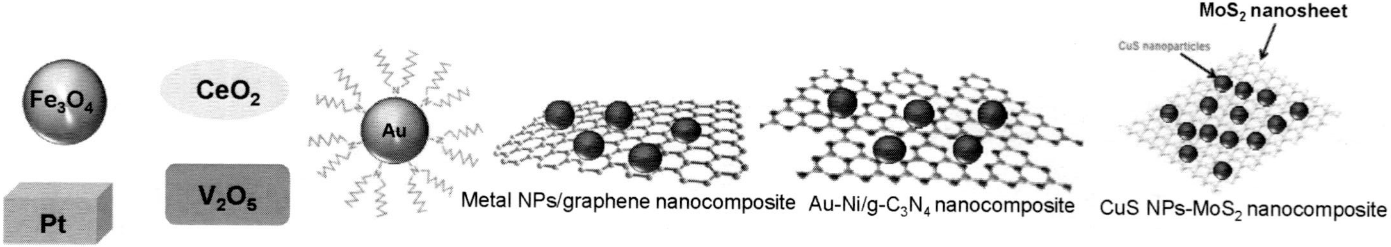

Figure 1. Examples of nanomaterials showing enzymatic features.

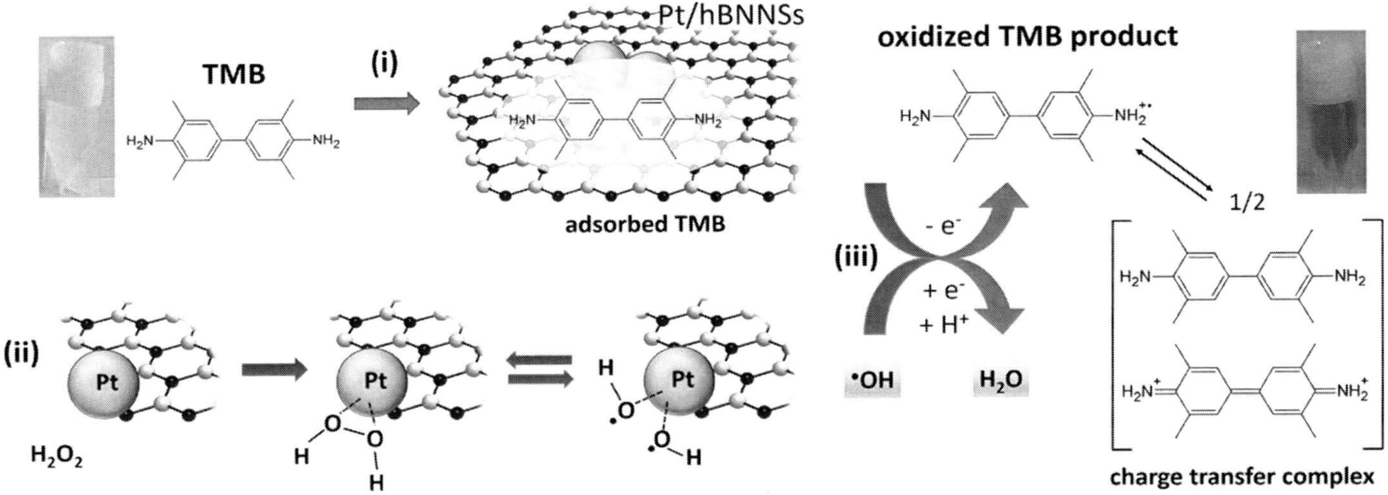

Reproduced with permission from reference [52].

Figure 2. Illustration of the plausible peroxidase mimic mechanism of the Pt/hBNNSs nanozyme. The conversion of the chromogenic substrate TMB to oxTMB plausibly took place in three steps: (i) Adsorption of TMB on the catalyst; (ii) Decomposition of H_2O_2 on the surface of Pt NPs; (iii) Diffusion of •OH radical species to oxidize TMB into the blue-colored oxTMB product through a one-electron process.

3.1. Overview of Peroxidase Mimics

Peroxidases represent a class of isoenzymes that commonly catalyze the decomposition of peroxides [30]. In this catalysis, peroxidases can convert H_2O_2 to water and oxygen. By scavenging H_2O_2, peroxidases play a significant role in various biological processes in almost all living organisms. The levels of peroxidases in the human body can directly or indirectly be used as biomarkers in various diseases. Natural peroxidases (e.g., Horseradish peroxidase (HRP)) possess a heme cofactor in their catalytic sites [31]. Because of their oxidative nature, peroxidase enzymes provide a wide variety of practical applications that include bioremediation of waste water, dye decolorization, transformation of pesticides, detection of antigens or antibodies, biosensing and diagnostics, etc. [32, 33].

In 2007, Yan and co-workers established the inherent peroxidase-mimicking catalytic features of Fe_3O_4 magnetic NPs by oxidizing colorless chromogenic substrates, namely, TMB, diazoaminobenzene, and OPD, to their oxidized colored products with the aid of H_2O_2 [34]. The rate of the reaction was evaluated by tracking the change in the absorbance intensity at different times. At a fixed catalyst concentration, kinetic studies were conducted by changing the concentration of a chromogenic substrate or the H_2O_2. Kinetic parameters such as the reaction velocity (v_{max}) and Michaelis–Menten constants (K_m) were determined to confirm the enzyme-like catalytic activity by Lineweaver–Burk Plots. The kinetics studies revealed a ping-pong mechanism for the nanozymatic catalysis of Fe_3O_4 NPs. The Michaelis–Menten constants suggested that the nanozyme had higher affinities to substrate TMB and lower affinities to substrate H_2O_2 than HRP. Based on the findings, the group of Yan [34] established Fe_3O_4 NPs as efficient artificial enzymes for various applications in the field of biotechnology, medicine, and environmental chemistry. Inspired by the work of Yan, peroxidase mimicking activities of a variety of nanomaterials have been investigated. For instance, Tremel and co-workers discovered the peroxidase mimicking activity of V_2O_5 nanowires and demonstrated their applicability in marine anti-biofouling [35].

Several researchers demonstrated the peroxidase mimicking catalytic activities of monometallic and multimetallic noble metal NPs such as Au, Ag, Pd, and Pt and used these nanomaterials for biosensors, antibiosis, and therapy [28, 36-41]. In 2010, the group of Qu demonstrated that both single-walled carbon nanotubes and GO possess temperature, pH, and H_2O_2 concentrationdependent catalytic activities similar to HRP [42, 43]. Das and co-workers developed various peroxidase mimic nanocomposite materials and applied them in numerous sensor applications [44-49]. For instance, they decorated Cu-Ag bimetallic NPs on rGO nanosheets and Au-Ni bimetallic NPs on doped graphitic carbon nitride nanosheets for colorimetric detection of glucose and ascorbic acid with high sensitivity [50, 51]. Recently Ivanova, Grayfer, Das, and co-workers established Pt NPs decorated on h-BN nanosheets (Pt/hBNNSs) as the colorimetric sensor for detecting the neurotransmitter compound 'dopamine' present in real blood serum samples [52]. Figure 2 depicts the plausible peroxidase mimic catalytic mechanism of the Pt/hBNNSs nanozyme in oxidizing colorless TMB to blue colored oxidized TMB (oxTMB) product with the aid of H_2O_2.

3.2. Overview of Oxidase Mimics

Oxidases are a class of enzymes that catalyze the oxidation of substrates with the aid of molecular O_2 into oxidized products, and O_2 is reduced to H_2O or H_2O_2. An important example of natural oxidase is cytochrome c oxidase which enables the body to utilize molecular O_2 to generate energy and in the respiratory electron transport chain of cells [53]. Another important example is the glucose oxidase enzyme (GOx) that catalyzes glucose oxidation to gluconolactone (which then hydrolyzes to gluconic acid) by molecular O_2 with the simultaneous formation of H_2O_2 [54].

In the last two decades, several nanomaterials have been reported to exhibit the catalytic properties of oxidases. Rossi and co-workers reported that carbon-supported Au NPs could catalyze D-glucose oxidation to D-gluconic acid under mild conditions at both controlled (7–9.5) and free pH values using molecular O_2 as the oxidant [55]. The same group further reported that unsupported Au NPs particles (average diameter 3.5 nm) also exhibited good GOx mimic activity for the aerobic oxidation of glucose [56]. The catalytic activity was inversely correlated to the particle diameter. Compared to naked Au NPs, the supported catalyst showed long-standing activity by preventing the aggregation of NPs.

Based on kinetic measurements, Rossi and co-workers proposed the mechanism of aerobic glucose oxidation by Au NPs under alkaline conditions [57, 58], presented in Figure 3.

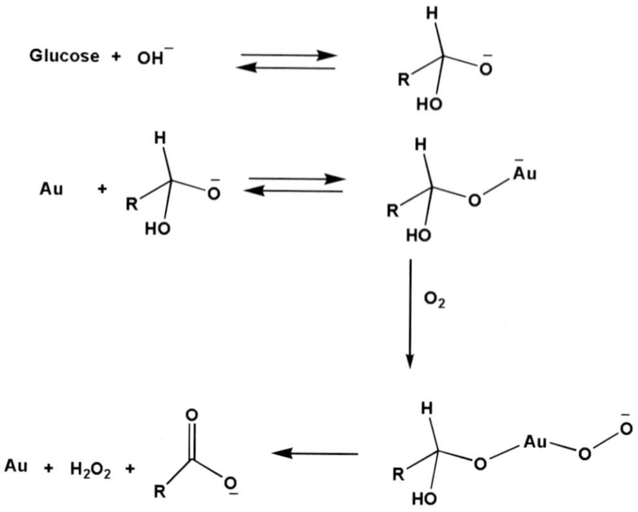

Figure 3. Plausible oxidase mimic mechanism of Au NPs in glucose oxidation as proposed by Rossi and co-workers [58].

First, the hydrated glucose anion gets adsorbed onto the surface of Au NPs, forming negatively charged gold species. Then the negatively charged Au NPs activate dissolved molecular O_2 to generate a dioxo Au intermediate. Finally, gluconic acid and H_2O_2 are released from the Au surface. After the discovery AuNPs as GOx mimics, other supported gold nanozymes, and Cu, Mo, Pt-based nanozymes have been widely explored as GOx mimics [28, 59, 60].

4. Sensing of Pesticides with Nanozymes

In the literature, the sensing of pesticides using nanozymes is performed by either single nanozyme sensors or integrated enzyme–nanozyme sensors. This book chapter covers a few salient examples of nanozymes as sensors for pesticides detection to give the readers a general overview of the broad research area. First, we discuss the examples of pesticides detection by single nanozyme sensors. Das and co-worker developed a Fe_3O_4-TiO_2/ rGO nanocomposite and employed it as a dual responsive nanozyme for the colorimetric detection of a harmful atrazine pesticide as well as photodegradation of atrazine in an aqueous solution [61].

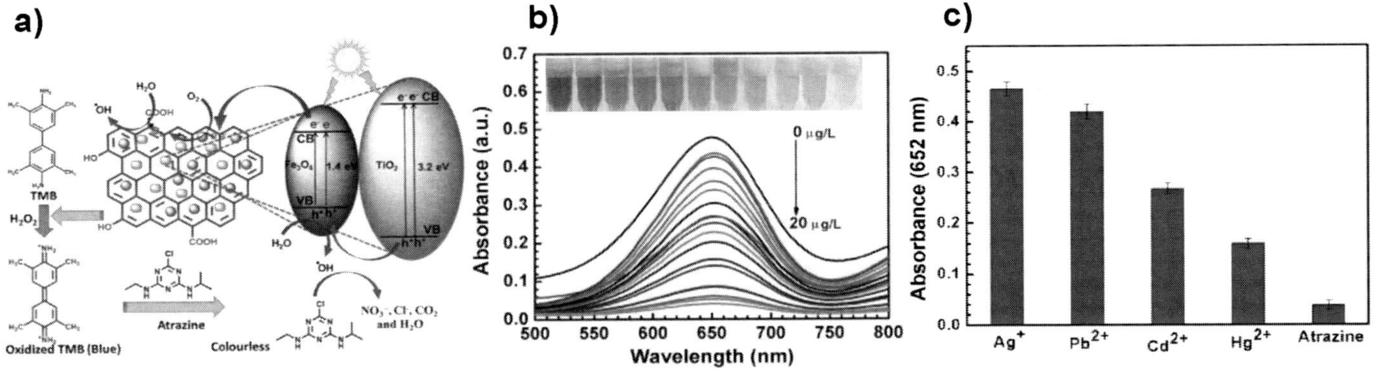

Reproduced with permission from reference [61].

Figure 4. a) Mechanism of peroxidase mimic sensing of atrazine pesticide and its photodegradation with Fe_3O_4-TiO_2/rGO nanozyme; b) Variation of absorbance intensity of oxTMB at 652 nm with the change in concentration of atrazine; c) Selectivity towards atrazine detection in comparison to interfering ions.

The nanocomposite (Fe_3O_4-TiO_2/rGO) was fabricated by a one-pot solvothermal method to simultaneously deposit magnetic Fe_3O_4 and semiconducting TiO_2 NPs onto the rGO surface. Transmission electron microscopic (TEM) measurements revealed that uniformly distributed Fe_3O_4 and TiO_2 NPs having an average size of 9 ± 0.2 nm were deposited on rGO nanosheets. The nanocomposite exhibited high intrinsic peroxidase mimic activity as confirmed by performing Michaelis–Menten kinetics in the oxidation of the chromogenic substrate TMB to the blue-colored oxTMB product with the aid of H_2O_2. In the presence of the atrazine pesticide, the oxidation of TMB to the blue-colored product (absorbance at 652 nm) is inhibited plausibly due to the noncovalent interactions between atrazine and TMB. Due to this inhibition, the nanocomposite was employed as an efficient sensor for the selective colorimetric detection of atrazine with a 2.98 μgmL^{-1} level of the limit of detection (LOD). The nanocomposite was utilized further for the 100% photocatalytic degradation of atrazine molecules into harmless products under the irradiation of natural solar light. Comparison with controlled catalyst samples confirmed the synergistic effect between Fe_3O_4 and TiO_2 NPs in the catalysis by the Fe_3O_4-TiO_2/rGO nanocomposite. Peroxidase mimic sensing of atrazine pesticide and its photocatalytic degradation with the Fe_3O_4-TiO_2/rGO nanozyme is shown in Figure 4. In a subsequent study, Das and co-workers developed a nanocomposite: polydopamine functionalized rGO decorated with magnetic Fe_3O_4 NPs and established it as an efficient nanozyme sensor for the colorimetric detection of harmful simazine pesticide with a LOD of 2.24 μM and as a photocatalyst for the sunlight-assisted degradation of the pesticide [62].

Wei, Guo, and co-workers developed colorimetric nanozyme sensor arrays using heteroatom-doped graphene as peroxidase mimics to detect and distinguish five aromatic pesticides [63]. Heteroatom-doped graphene materials catalyzed the oxidation of TMB to blue-colored oxTMB with the aid of H_2O_2. In the presence of aromatic pesticides, the active sites of nanozymes were masked to different degrees due to the adsorption of pesticides on the graphene surfaces, lowering the peroxidase-mimicking activities of the nanozymes. Heteroatom-doped graphene materials were synthesized by the high-temperature heat treatment of mixtures of GO with urea and benzyl disulphide. The principle of pesticides detection and discrimination by the sensor arrays is illustrated in Figure 5.

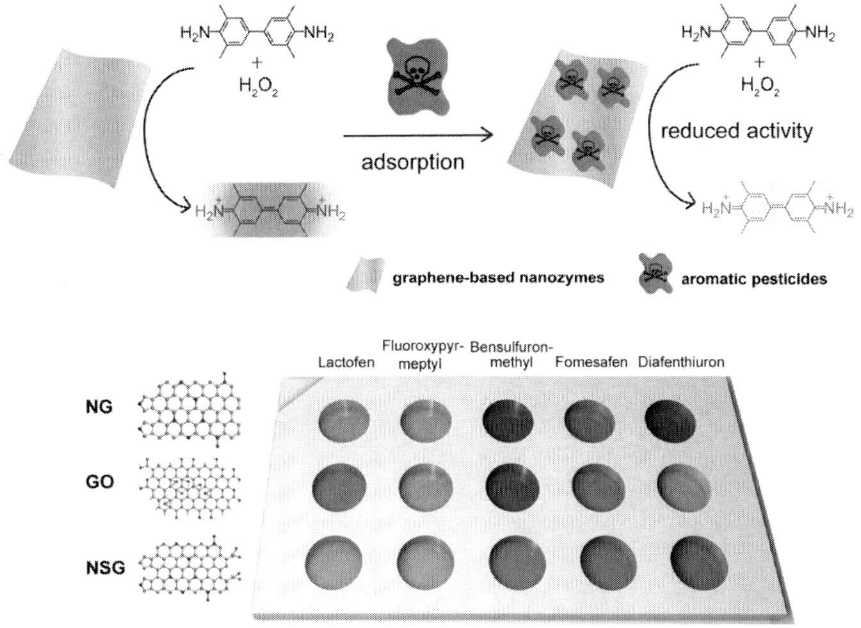

Reproduced with permission from reference [63].

Figure 5. Pictorial representation of the functioning of heteroatom-doped graphene-based nanozyme sensor arrays for detecting and discriminating different aromatic pesticides.

Based on the enzymatic activity inhibition phenomenon, five pesticides, i.e., bensulfuron-methyl, diafenthiuron, fluoroxypyr-meptyl, fomesafen, and lactofen, were successfully discriminated in the wide range 5 to 500 μM of pesticides concentration. The sensor arrays could also effectively discriminate different concentrations of individual pesticides and different ratios of two pesticides when mixed together. The sensor arrays were successfully demonstrated to discriminate the pesticides present in soil samples.

Pan and co-workers developed a colorimetric assay for detecting the malathion pesticide [64] using the oxidase-mimetic property of Ag_3PO_4/UiO-66 nanozyme. The nanozyme catalyzes the oxidation of peroxidase substrate TMB to blue-colored oxTMB. The oxidase-like activity of the nanozyme could be inhibited by the pesticide as it forms Ag-S bonds with the nanozyme. Upon adding the pesticide, the amount of oxTMB product generated becomes lower, which could be used to detect the malathion pesticide in a wide concentration range of 0.0083–5.333 μgmL^{-1} with a detection limit of 7.5 ng/mL. Moreover, the authors fabricated a smartphone sensing system based on chromogenic hydrogel material for on-site analysis of the malathion pesticide.

Lawati and co-workers developed paper-based chemiluminescence (CL) device to determine total phenolic content in food samples [65]. Note that phenolic compounds are also important elements of some pesticides. The paper-based CL sensor utilizes Co metal-organic framework (CoMOF) to decompose H_2O_2 to •OH radicals. Thus formed •OH radicals react with Rhodamine B molecules, located in the nanopores of CoMOF (R@CoMOF), generating an intense CL emission signal. When phenolic compounds were added to the paper containing the CL reagents, the CL intensity of the device significantly decreased due to the interference of phenolic compounds in the process of •OH radical production (Figure 6). This CL intensity inhibition effect was applied to fabricate a simple analytical assay to detect and quantify phenolic compounds. After careful optimization, the analytical device detected gallic acid, quercetin, catechin, kaempferol, and caffeic acid with LODs of 0.98, 1.36, 1.48, 1.81, and 2.55 ng mL^{-1}. Furthermore, the device was successfully employed to determine the total antioxidant capacity of real food products.

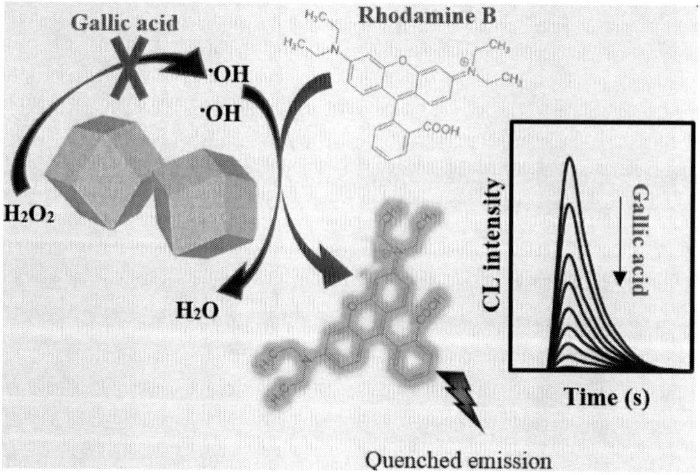

Reproduced with permission from reference [65].

Figure 6. Principle of functioning of paper-based chemiluminescence device in determining the total phenolic content in food products.

In the next succeeding section, we describe pesticides detection by integrated enzyme–nanozyme sensors. Habibi and co-workers developed a nanozyme by depositing magnetic Fe_3O_4 NPs into ZIF-8 and used the nanocomposite Fe_3O_4NPs@ZIF-8 for the sensitive biosensing of organophosphate (OP) pesticides (Figure 7).

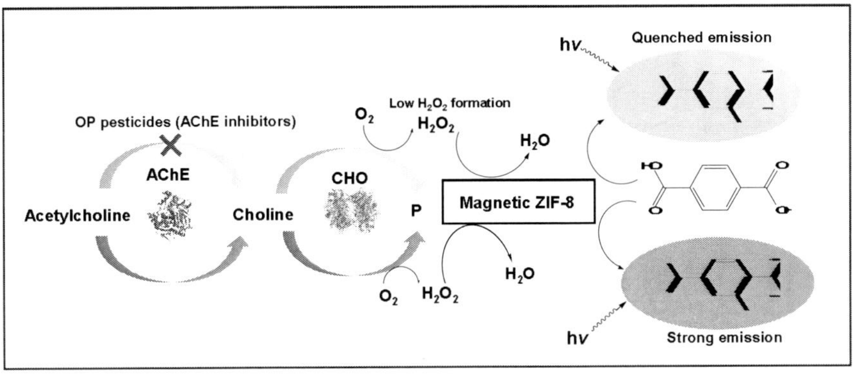

Figure 7. Illustration of fluorescence detection of organophosphate pesticides using the acetylcholinesterase/choline oxidase/Fe_3O_4NPs@ZIF-8 nanozyme integrated sensor developed by Habibi and co-workers [66].

The detection method employed a fluorometric assay by integrating the acetylcholinesterase (AChE) inhibited enzymatic activity and the peroxidase-like activity of Fe_3O_4NPs@ZIF-8 nanozyme to detect OPs [66]. The sensor combines AChE and choline oxidase (CHO) enzymes with Fe_3O_4NPs@ZIF-8 nanozyme. The peroxidase mimic Fe_3O_4NPs@ZIF-8 nanozyme decomposes H_2O_2 generated during the enzymatic reactions of acetylcholine to •OH radicals. The radicals, in turn, oxidize the peroxidase substrate terephthalic acid to generate a fluorescent product with a high detectable signal. When an OP pesticide is added, due to its inhibition effect on the activity of the AChE enzyme, the production rate of H_2O_2 decreases, leading to a decrease in the generated fluorescent signal proportional to the pesticide concentration. The limit of detection was reported to be 0.2 nM. The developed sensor was successfully utilized to detect and quantify the diazinon pesticide in water and fruit juices.

Lin and co-workers developed a unique colorimetric sensing method for quantitatively detecting the enzymatic activity of AChE and OP pesticides as AChE inhibitors. They employed MnO_2 nanosheets as the oxidase-mimicking nanozyme that directly oxidizes the TMB substrate to blue-colored oxTMB product (absorbance at 652 nm) without H_2O_2 [67]. When AChE and acetylthiocholine were introduced to a solution containing MnO_2 nanozyme and TMB, a decrease of solution absorbance at 652 nm was observed. The decrease of absorbance was due to the fact that AChE catalyzed the hydrolysis of acetylthiocholine to produce thiocholine that triggered the decomposition of MnO_2 nanosheets, causing an inhibition of its peroxidase-like activity. In the presence of an OP pesticide, the activity of AChE was suppressed, which prevented the decomposition of MnO_2 nanosheets and resulted in an increase of solution absorbance at 652 nm. The authors reported that under optimized conditions, the colorimetric method exhibited a high sensitive to AChE and the paraoxon pesticide in the wide ranges of 0.1-15 $mUmL^{-1}$ and 0.001-0.1 μgmL^{-1}. The limits of detection of AChE and paraoxon were reported to be 35 μUmL^{-1} and 1.0 $ngmL^{-1}$, respectively. Moreover, the MnO_2 nanozyme-based sensor was used to design test strips for rapid naked-eye detection of AChE and AChE inhibitors in highly promising fashion.

Conclusion

The uncontrolled use of toxic pesticides in the agricultural sector has caused immense threats to the environment and human health, which urgently requires developing effective methods and tools for quick detection of pesticides in food, water, and soil samples. Nanozymes having the advantages of affordable costs, tolerance to harsh conditions, long-standing stability, and scalable production find great potential to be used as sensors for the rapid detection of pesticides.

In this chapter, we have discussed developments of nanozymes as biosensors for detecting pesticides and their related biomarkers. We have given a general understanding of the principles of nanozymes with a particular emphasis on peroxidase and oxidase mimics. A few salient examples of the application of nanozymes as sensors for detection of pesticides have been discussed. Although great advancements have been made in pesticides sensing using diverse nanomaterials, a few challenges remain in the emerging stage, requiring further efforts for practical implications of nanozyme-based pesticide sensors. A few important research targets in this area can be identified as follows. Many pesticide detection sensors combine one or more bio-enzyme(s) with a nanozyme. Therefore, research on developing sensor systems composed of a cascade of nanomaterials for pesticides sensing is highly important.

In the case of real samples, as many interfering chemicals may present along with the analyte, selectivity is a major problem for nanozyme-based sensors. Therefore, future research work on the fabrication of more nanozyme-based sensor arrays to identify and differentiate multiple targets is also important. Advanced detection techniques such as electrochemical and SERS need to be equipped with nanozyme catalysis to improve the sensitivity of nanozymes towards pesticides. Even though large number of nanozyme-based sensing methods have been explored and verified in the laboratories for pesticides detection with excellent performance, their feasibility in real environmental samples still needs further validation.

For practical applications, it is important to standardize fabrication and utilization methods of nanozymes via some analytical devices such as paper-based analytical devices. These analytical devices must be inexpensive, easy-to-use, portable and highly sensitive for the rapid detection of pesticides.

Acknowledgments

The authors are thankful to the Director, CSIR-NEIST, Jorhat, India for his permission to carry out the work. The authors also acknowledge CSIR for financial support (Project No. MLP 1013 and OLP 2074).

References

[1] Aktar, W., D. Sengupta, A. Chowdhury, Impact of pesticides use in agriculture: their benefits and hazards, *Interdisciplinary Toxicology*, 2 (2009) 1-12.

[2] Carvalho, F. P. Pesticides, environment, and food safety, *Food and Energy Security,* 6 (2017) 48-60.

[3] Mogul, M. G., H. Akin, N. Hasirci, D. J. Trantolo, J. D. Gresser, D. L. Wise, Controlled release of biologically active agents for purposes of agricultural crop management, *Resources, Conservation and Recycling*, 16 (1996) 289-320.

[4] Damalas, C. A., I. G. Eleftherohorinos, Pesticide Exposure, Safety Issues, and Risk Assessment Indicators, *International Journal of Environmental Research and Public Health*, 8 (2011) 1402-1419.

[5] Kim, K. H., E. Kabir, S. A. Jahan, Exposure to pesticides and the associated human health effects, *Science of The Total Environment*, 575 (2017) 525-535.

[6] McKnight, U. S., J. J. Rasmussen, B. Kronvang, P. J. Binning, P. L. Bjerg, Sources, occurrence and predicted aquatic impact of legacy and contemporary pesticides in streams, *Environmental Pollution*, 200 (2015) 64-76.

[7] Hernández, F., J. V. Sancho, O. J. Pozo, Critical review of the application of liquid chromatography/mass spectrometry to the determination of pesticide residues in biological samples, *Analytical and Bioanalytical Chemistry*, 382 (2005) 934-946.

[8] Gabaldón, J. A., A. Maquieira, R. Puchades, Development of a simple extraction procedure for chlorpyrifos determination in food samples by immunoassay, *Talanta,* 71 (2007) 1001-1010.
[9] Wang, P., L. Wu, Z. Lu, Q. Li, W. Yin, F. Ding, H. Han, Gecko-Inspired Nanotentacle Surface-Enhanced Raman Spectroscopy Substrate for Sampling and Reliable Detection of Pesticide Residues in Fruits and Vegetables, *Analytical Chemistry,* 89 (2017) 2424-2431.
[10] Liu, S., Z. Zheng, F. Wei, Y. Ren, W. Gui, H. Wu, G. Zhu, Simultaneous Determination of Seven Neonicotinoid Pesticide Residues in Food by Ultraperformance Liquid Chromatography Tandem Mass Spectrometry, *Journal of Agricultural and Food Chemistry*, 58 (2010) 3271-3278.
[11] Qian, G., L. Wang, Y. Wu, Q. Zhang, Q. Sun, Y. Liu, F. Liu, A monoclonal antibody-based sensitive enzyme-linked immunosorbent assay (ELISA) for the analysis of the organophosphorous pesticides chlorpyrifos-methyl in real samples, *Food Chemistry*, 117 (2009) 364-370.
[12] Sporty, J. L. S., S. W. Lemire, E. M. Jakubowski, J. A. Renner, R. A. Evans, R. F. Williams, J. G. Schmidt, M. J. V. D. Schans, D. Noort, R. C. Johnson, Immunomagnetic Separation and Quantification of Butyrylcholinesterase Nerve Agent Adducts in Human Serum, *Analytical Chemistry*, 82 (2010) 6593-6600.
[13] Verma, N., A. Bhardwaj, Biosensor Technology for Pesticides—A review, *Applied Biochemistry and Biotechnology*, 175 (2015) 3093-3119.
[14] Gong, J., Z. Guan, D. Song, Biosensor based on acetylcholinesterase immobilized onto layered double hydroxides for flow injection/amperometric detection of organophosphate pesticides, *Biosensors and Bioelectronics*, 39 (2013) 320-323.
[15] Lyagin, I. V., E. N. Efremenko, S. D. Varfolomeev, Enzymatic biosensors for determination of pesticides, *Russian Chemical Reviews*, 86 (2017) 339-355.
[16] Sun, J., L. Guo, Y. Bao, J. Xie, A simple, label-free AuNPs-based colorimetric ultrasensitive detection of nerve agents and highly toxic organophosphate pesticide, *Biosensors and Bioelectronics*, 28 (2011) 152-157.
[17] Liu D., W. Chen, J. Wei, X. Li, Z. Wang, X. Jiang, A Highly Sensitive, Dual-Readout Assay Based on Gold Nanoparticles for Organophosphorus and Carbamate Pesticides, *Analytical Chemistry*, 84 (2012) 4185-4191.
[18] Yi, Y., G. Zhu, C. Liu, Y. Huang, Y. Zhang, H. Li, J. Zhao, S. Yao, A Label-Free Silicon Quantum Dots-Based Photoluminescence Sensor for Ultrasensitive Detection of Pesticides, *Analytical Chemistry*, 85 (2013) 11464-11470.
[19] Zhu, H., P. Liu, L. Xu, X. Li, P. Hu, B. Liu, J. Pan, F. Yang, X. Niu, Nanozyme-Participated Biosensing of Pesticides and Cholinesterases: A Critical Review, *Biosensors,* 11 (2021) 382.
[20] Naveen Prasad, S., V. Bansal, R. Ramanathan, Detection of pesticides using nanozymes: Trends, challenges and outlook, *TrAC Trends in Analytical Chemistry,* 144 (2021) 116429.
[21] Huang, Y., X. Mu, J. Wang, Y. Wang, J. Xie, R. Ying, E. Su, The recent development of nanozymes for food quality and safety detection, *Journal of Materials Chemistry B*, 10 (2022) 1359-1368.

[22] Wang, H., K. Wan, X. Shi, Recent Advances in Nanozyme Research, *Advanced Materials*, 31 (2019) 1805368.

[23] Jayaraj, R., P. Megha, P. Sreedev, Review Article. Organochlorine pesticides, their toxic effects on living organisms and their fate in the environment, *Interdisciplinary Toxicology,* 9 (2016) 90-100.

[24] Mearns, J., J. Dunn, P. R. Lees-Haley, Psychological effects of organophosphate pesticides: A review and call for research by psychologists, *Journal of Clinical Psychology,* 50 (1994) 286-294.

[25] I. Dhouib, M. Jallouli, A. Annabi, S. Marzouki, N. Gharbi, S. Elfazaa, M. M. Lasram, From immunotoxicity to carcinogenicity: the effects of carbamate pesticides on the immune system, *Environmental Science and Pollution Research*, 23 (2016) 9448-9458.

[26] Holt, J. S., Herbicides, in: S. A. Levin (Ed.) *Encyclopedia of Biodiversity* (Second Edition), Academic Press, Waltham, 2013, pp. 87-95.

[27] Wei, H., E. Wang, Nanomaterials with enzyme-like characteristics (nanozymes): next-generation artificial enzymes, *Chemical Society Reviews*, 42 (2013) 6060-6093.

[28] Wu, J., X. Wang, Q. Wang, Z. Lou, S. Li, Y. Zhu, L. Qin, H. Wei, Nanomaterials with enzyme-like characteristics (nanozymes): next-generation artificial enzymes (II), *Chemical Society Reviews*, 48 (2019) 1004-1076.

[29] Huang, Y., J. Ren, X. Qu, Nanozymes: Classification, Catalytic Mechanisms, Activity Regulation, and Applications, *Chemical Reviews*, 119 (2019) 4357-4412.

[30] Attar, F., M. G. Shahpar, B. Rasti, M. Sharifi, A. A. Saboury, S. M. Rezayat, M. Falahati, Nanozymes with intrinsic peroxidase-like activities, *Journal of Molecular Liquids,* 278 (2019) 130-144.

[31] Veitch, N. C. Horseradish peroxidase: a modern view of a classic enzyme, *Phytochemistry,* 65 (2004) 249-259.

[32] Regalado, C., B. E. García-Almendárez, M. A. Duarte-Vázquez, Biotechnological applications of peroxidases, *Phytochemistry Reviews,* 3 (2004) 243-256.

[33] Hamid, M., R. Khalil ur, Potential applications of peroxidases, *Food Chemistry*, 115 (2009) 1177-1186.

[34] Gao, L., J. Zhuang, L. Nie, J. Zhang, Y. Zhang, N. Gu, T. Wang, J. Feng, D. Yang, S. Perrett, X. Yan, Intrinsic peroxidase-like activity of ferromagnetic nanoparticles, *Nature Nanotechnology*, 2 (2007) 577-583.

[35] André, R., F. Natálio, M. Humanes, J. Leppin, K. Heinze, R. Wever, H. C. Schröder, W. E. G. Müller, W. Tremel, V2O5 Nanowires with an Intrinsic Peroxidase-Like Activity, *Advanced Functional Materials*, 21 (2011) 501-509.

[36] Wu, Y. S., F. F. Huang, Y. W. Lin, Fluorescent Detection of Lead in Environmental Water and Urine Samples Using Enzyme Mimics of Catechin-Synthesized Au Nanoparticles, *ACS Applied Materials & Interfaces*, 5 (2013) 1503-1509.

[37] Chen, L., L. Sha, Y. Qiu, G. Wang, H. Jiang, X. Zhang, An amplified electrochemical aptasensor based on hybridization chain reactions and catalysis of silver nanoclusters, *Nanoscale,* 7 (2015) 3300-3308.

[38] Li, W., B. Chen, H. Zhang, Y. Sun, J. Wang, J. Zhang, Y. Fu, BSA-stabilized Pt nanozyme for peroxidase mimetics and its application on colorimetric detection of mercury(II) ions, *Biosensors and Bioelectronics*, 66 (2015) 251-258.

[39] Ge, S., F. Liu, W. Liu, M. Yan, X. Song, J. Yu, Colorimetric assay of K-562 cells based on folic acid-conjugated porous bimetallic Pd@Au nanoparticles for point-of-care testing, *Chemical Communications*, 50 (2014) 475-477.

[40] Zheng, C., A. X. Zheng, B. Liu, X. L. Zhang, Y. He, J. Li, H. H. Yang, G. Chen, One-pot synthesized DNA-templated Ag/Pt bimetallic nanoclusters as peroxidase mimics for colorimetric detection of thrombin, *Chemical Communications*, 50 (2014) 13103-13106.

[41] Jiang, T., Y. Song, T. Wei, H. Li, D. Du, M. J. Zhu, Y. Lin, Sensitive detection of Escherichia coli O157:H7 using Pt–Au bimetal nanoparticles with peroxidase-like amplification, *Biosensors and Bioelectronics*, 77 (2016) 687-694.

[42] Song, Y., K. Qu, C. Zhao, J. Ren, X. Qu, Graphene Oxide: Intrinsic Peroxidase Catalytic Activity and Its Application to Glucose Detection, *Advanced Materials*, 22 (2010) 2206-2210.

[43] Song, Y., X. Wang, C. Zhao, K. Qu, J. Ren, X. Qu, Label-Free Colorimetric Detection of Single Nucleotide Polymorphism by Using Single-Walled Carbon Nanotube Intrinsic Peroxidase-Like Activity, *Chemistry – A European Journal*, 16 (2010) 3617-3621.

[44] Borthakur, P., P. K. Boruah, M. R. Das, S. B. Artemkina, P. A. Poltarak, V. E. Fedorov, Metal free MoS2 2D sheets as a peroxidase enzyme and visible-light-induced photocatalyst towards detection and reduction of Cr(vi) ions, *New Journal of Chemistry,* 42 (2018) 16919-16929.

[45] Das, P., P. Borthakur, P. K. Boruah, M. R. Das, Peroxidase Mimic Activity of Au–Ag/l-Cys-rGO Nanozyme toward Detection of Cr(VI) Ion in Water: Role of 3,3′,5,5′-Tetramethylbenzidine Adsorption, *Journal of Chemical & Engineering Data,* 64 (2019) 4977-4990.

[46] Borthakur, P., M. R. Das, S. Szunerits, R. Boukherroub, CuS Decorated Functionalized Reduced Graphene Oxide: A Dual Responsive Nanozyme for Selective Detection and Photoreduction of Cr(VI) in an Aqueous Medium, *ACS Sustainable Chemistry & Engineering,* 7 (2019) 16131-16143.

[47] Borthakur, P., P. K. Boruah, M. R. Das, CuS and NiS Nanoparticle-Decorated Porous-Reduced Graphene Oxide Sheets as Efficient Peroxidase Nanozymes for Easy Colorimetric Detection of Hg(II) Ions in a Water Medium and Using a Paper Strip, *ACS Sustainable Chemistry & Engineering,* 9 (2021) 13245-13255.

[48] Borthakur, P., G. Darabdhara, M. R. Das, R. Boukherroub, S. Szunerits, Solvothermal synthesis of CoS/reduced porous graphene oxide nanocomposite for selective colorimetric detection of Hg(II) ion in aqueous medium, *Sensors and Actuators B: Chemical,* 244 (2017) 684-692.

[49] Borthakur, P., P. K. Boruah, M. R. Das, Facile synthesis of CuS nanoparticles on two-dimensional nanosheets as efficient artificial nanozyme for detection of Ibuprofen in water, *Journal of Environmental Chemical Engineering,* 9 (2021) 104635.

[50] Darabdhara, G., B. Sharma, M. R. Das, R. Boukherroub, S. Szunerits, Cu-Ag bimetallic nanoparticles on reduced graphene oxide nanosheets as peroxidase mimic for glucose and ascorbic acid detection, *Sensors and Actuators B: Chemical*, 238 (2017) 842-851.

[51] Darabdhara, G., J. Bordoloi, P. Manna, M. R. Das, Biocompatible bimetallic Au-Ni doped graphitic carbon nitride sheets: A novel peroxidase-mimicking artificial enzyme for rapid and highly sensitive colorimetric detection of glucose, *Sensors and Actuators B: Chemical,* 285 (2019) 277-290.

[52] Ivanova, M. N., E. D. Grayfer, E. E. Plotnikova, L. S. Kibis, G. Darabdhara, P. K. Boruah, M. R. Das, V. E. Fedorov, Pt-Decorated Boron Nitride Nanosheets as Artificial Nanozyme for Detection of Dopamine, *ACS Applied Materials & Interfaces,* 11 (2019) 22102-22112.

[53] Michel, H., J. Behr, a. A. Harrenga, A. Kannt, CYTOCHROME C OXIDASE: Structure and Spectroscopy, *Annual Review of Biophysics and Biomolecular Structure,* 27 (1998) 329-356.

[54] Wilson, R., A. P. F. Turner, Glucose oxidase: an ideal enzyme, *Biosensors and Bioelectronics*, 7 (1992) 165-185.

[55] Biella, S., L. Prati, M. Rossi, Selective Oxidation of D-Glucose on Gold Catalyst, *Journal of Catalysis*, 206 (2002) 242-247.

[56] Comotti, M., C. Della Pina, R. Matarrese, M. Rossi, The Catalytic Activity of "Naked" Gold Particles, *Angewandte Chemie International Edition*, 43 (2004) 5812-5815.

[57] Beltrame, P., M. Comotti, C. Della Pina, M. Rossi, Aerobic oxidation of glucose: II. Catalysis by colloidal gold, *Applied Catalysis A: General*, 297 (2006) 1-7.

[58] Comotti, M., C. Della Pina, E. Falletta, M. Rossi, Aerobic Oxidation of Glucose with Gold Catalyst: Hydrogen Peroxide as Intermediate and Reagent, *Advanced Synthesis & Catalysis*, 348 (2006) 313-316.

[59] Chen, J., Q. Ma, M. Li, D. Chao, L. Huang, W. Wu, Y. Fang, S. Dong, Glucose-oxidase like catalytic mechanism of noble metal nanozymes, *Nature Communications,* 12 (2021) 3375.

[60] Chong, Y., Q. Liu, C. Ge, Advances in oxidase-mimicking nanozymes: Classification, activity regulation and biomedical applications, *Nano Today,* 37 (2021) 101076.

[61] Boruah, P. K., M. R. Das, Dual responsive magnetic Fe_3O_4-TiO_2/graphene nanocomposite as an artificial nanozyme for the colorimetric detection and photodegradation of pesticide in an aqueous medium, *Journal of Hazardous Materials*, 385 (2020) 121516.

[62] Boruah, P. K., G. Darabdhara, M. R. Das, Polydopamine functionalized graphene sheets decorated with magnetic metal oxide nanoparticles as efficient nanozyme for the detection and degradation of harmful triazine pesticides, *Chemosphere,* 268 (2021) 129328.

[63] Zhu, Y., J. Wu, L. Han, X. Wang, W. Li, H. Guo, H. Wei, Nanozyme Sensor Arrays Based on Heteroatom-Doped Graphene for Detecting Pesticides, *Analytical Chemistry,* 92 (2020) 7444-7452.

[64] Liu, P., X. Li, X. Xu, X. Niu, M. Wang, H. Zhu, J. Pan, Analyte-triggered oxidase-mimetic activity loss of Ag_3PO_4/UiO-66 enables colorimetric detection of malathion completely free from bioenzymes, *Sensors and Actuators B: Chemical*, 338 (2021) 129866.

[65] Hassanzadeh, J., H. A. J. Al Lawati, I. Al Lawati, Metal–Organic Framework Loaded by Rhodamine B As a Novel Chemiluminescence System for the Paper-Based Analytical Devices and Its Application for Total Phenolic Content Determination in Food Samples, *Analytical Chemistry,* 91 (2019) 10631-10639.

[66] Bagheri, N., A. Khataee, J. Hassanzadeh, B. Habibi, Sensitive biosensing of organophosphate pesticides using enzyme mimics of magnetic ZIF-8, *Spectrochimica Acta Part A: Molecular and Biomolecular Spectroscopy*, 209 (2019) 118-125.

[67] Yan, X., Y. Song, X. Wu, C. Zhu, X. Su, D. Du, Y. Lin, Oxidase-mimicking activity of ultrathin MnO_2 nanosheets in colorimetric assay of acetylcholinesterase activity, *Nanoscale,* 9 (2017) 2317-2323.

Chapter 3

Nanozymes for Organic Pollutant Detection

Adya Jain*
Department of Chemistry,
MRK Educational Institutions, Rewari, Haryana, India

Abstract

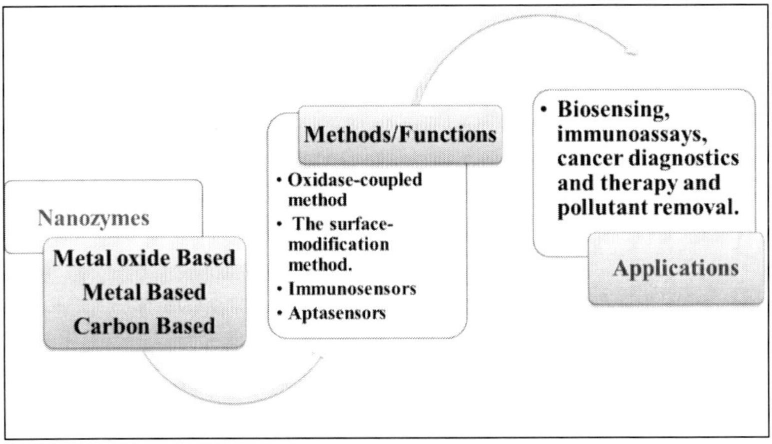

Figure 1. Graphical abstract.

Nanozymes are nanomaterials principally inorganic nanoparticles imitating natural enzymatic activity and catalyzed for environmental pollutant monitoring and remediation purposes such as removal of

* Corresponding Author's Email: adyajain02@gmail.com.

In: Emerging Environmental Applications of Nanozymes
Editors: Seema Nara and Smriti Singh
ISBN: 979-8-88697-552-9

organic pollutants which include pesticides, antibiotics, phenols, dyes, etc.

These nanozymes are economically sustainable as they offer effectiveness, reusability, selectivity, etc. also used as sensors for detection of organic compounds quantitatively as well as qualitatively. Typically, detection and their modulation by means of nanozymes is reviewed along with their specific strategies and mechanisms are broadly classified. No review focuses on the degradation of organic pollutant specifically by nanozymes.

Keywords: nanozymes, nanoparticles, enzymatic, organic pollutant, sensors, mechanisms

1. Introduction

1.1. Nanozymes

Nanozymes were initially recognized by Scrimin and Pasquato et al. having ribonuclease-like activity in thiol monolayer protected gold clusters [1].

Artificial enzymes were established long before comprising of molecules mimicking enzymatic activities. Nanozymes are actually artificial enzymes chemically composed of nanostructured materials, i.e., fullerenes, nanoparticles, nanosheets, nanocomposites, etc. Nanozymes have been thoroughly studied in wide applications such as biosensing, immunoassays, cancer diagnostics and therapy and pollutant removal. Nanoparticles and nanocomposites are mostly studied as catalysts due to high surface area to volume ratio, exceptional optical and electrical properties, and outstanding stability thus successfully employed for the degradation of noxious organic dyes. These promising artificial enzymes are stable structures with adjustable activities, high-efficiencies, low costs, large-scale production and recyclability [2, 3].

1.2. Organic Pollutants

Organic pollutants are organic compounds exceeding their permissible limits thus causing toxicity and diseases such as petroleum hydrocarbons, detergents, volatile organic solvents, pesticides plastics and dyes are the vivacious sources of these organic compounds. Majorly the POPs, i.e., persistent and poisonous organic compounds cause huge and long term effects on living-beings. Persistent organic pollutants include antibiotics, phenolic compounds, pesticides, dyes and organophosphorus compounds (OPPs, chemical warfare nerve agents and flame retardants) [4]. Phenolic compounds, predominantly chlorophenols, have been extensively used in pesticides, dyes, in various chemical reactions particularly in synthetic industries [5-7]. Therefore, an emerging aid for its removal is an urgent requirement.

1.3. Classification of Nanozymes

Nanozymes fundamentally work on two basic mechanisms, i.e., oxidase-coupled method and the surface-modification method. In oxidase-coupled method, nanozymes act as peroxidase like enzymes and catalyzes the oxidation of colorimetric substrates with the resulting H_2O_2. While surface-modification method is based on antigen-antibody interactions in which antibody is located at the surface of nanozyme and targeted for specific antigens [8].

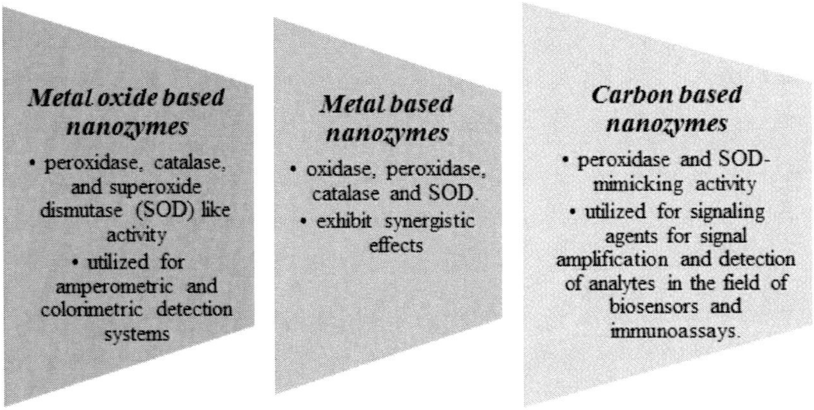

Figure 2. Classification of Nanozymes on the basis of composition.

It has also been reported that the nature of nanozyme changes on changing the acidic and basic conditions [9]. Nitrogen@carbon nanospheres (N-PCNSs) mimics four different types of enzyme-like activities, i.e., peroxidase, oxidase, catalase and superoxide dismutase (SOD)-like activities on varying pH and physiological conditions. Under acidic conditions, Oxidase and peroxidase like activities have a tendency to catalyze reactions analogous to the oxygen reduction reaction (ORR), on the other hand at basic conditions, catalase- and SOD-like activities have a tendency to perform on reversible reactions analogous to the oxygen evolution reaction (OER) [10].

Out of six major classes (i.e., oxidoreductases, transferases, hydrolases, lyases, isomerases and ligases) of enzymes nanozymes normally act as oxidoreductases and hydrolases catalyst. Oxidoreductases catalyse oxidations and reductions in which hydrogen or oxygen atoms or electrons are transferred between molecules such as oxidases (including laccases), superoxide dismutases, peroxidases and catalases while Hydrolases catalyse the hydrolysis of various bonds and include enzymes such as phosphatase, nuclease, protease and peptidase [11].

1.4. Degradation of Organic Pollutants via Nanozymes

Borthakur et al. reported the degradation of HQ (hydroquinone) in the presence of CuS-MoS_2 [12]. CuS nanoparticles decorated MoS_2 nanosheets shows photocatalytic and peroxidase-like activity. This nanozyme shows 83% HQ decomposition efficiency achieved in 240 min. Although variety of nanocomposites also have been reported for hydroquinone degradation activity such as MoS_2-Pt_3Au_1 [13], Au/TiO_2/RGO [14], TiO_2/AC [15], 3D porous ZnO/$ZnFe_2O_4$/graphene foam (3D ZZFO/GF) [16], Fe loaded polystyrene-divinyl benzene polymeric hybrid ion exchange (HIX) resin [17]. To investigate the peroxidase like activity the nanozyme has been treated with H_2O_2, thus blue colored oxidized compound was obtained. Further analyzed and confirmed by UV-Visible spectrophotometer. To optimize the conditions varying pH, catalytic concentration and time were also studied. Maximum absorption was found at pH 4 and temperature at 35°C. The fadedness of the blue colour with concentration change indicated the potentiality of the nanozyme. PXRD, TEM, BET and XPS analysis were used to determine the structure of nanozyme. Lower values of Michaelis-Menten constant (Km) indicates greater affinity of nanozymes towards degradation of HQ. This work has found to lead a new path for the utilization of semiconductor-based heterogeneous systems in the potential application of the environmental remediation process.

Ding et al. developed a rapid methodology for detecting melamine which is a toxic organic nitrogenous compound and illegally inoculated in dairy products by the help of Nanocomposite-entrapping MNPs [18]. ABTS has been used as colorimetric substrates. The catalytic oxidation of ABTS is inhibited by melamine in the presence of H_2O_2, because it reacts with H_2O_2 to form an additional compound. As a result, the intensity of the ABTS color signal was dependent on the concentration of melamine. Thus, a simple colorimetric system using MNPs can be established for naked eye detection of excess concentrations of melamine in dairy products. Nanocomposite-entrapping MNPs and oxidase in mesoporous carbon were used to detect several phenol compound amperometrically, such as phenol, cresol and cathechol [19]. These phenol compounds produced a concentration dependent increase of cathodic current in this system, which may have great potential in the field of environmental monitoring.

Various treatment methods also don't have a recognizable efficiency for removal organic pollutants in waste water. Nanozyme-based methods have been found to have distinct advantages influential, high thermal stability (5-90ºC), economical, reusability/recyclable with 100% efficiency and simple method for degradation and mineralization of organic dyes from industrial processes. Most significantly, MNPs mimicking as peroxidase like activity have been investigated for degradation of organic pollutants [20], methylene blue [21], phenol and rhodamine B. MNPs-H_2O_2 could eradicate 85% of phenol from aqueous solution within three hours [22]. Removal of methylene blue by MNPs-H_2O_2 coupled method has also been successfully performed by Jiang et al. [23]. It was observed that 96% of methylene blue was degraded in 15 minutes at optimized condition.

Gao et al. reported that MNPs-H_2O_2 system for the degradation of biofilm along with its bacteria [24]. These biofilm and bacteria are major cause of nosocomial infection [25]. The free hydroxyl radicals formed by MNP catalysis of H_2O_2 degrades the oxidative cleavage of biofilm components (nucleic acids, proteins, and polysaccharides) along with killing the bacteria.

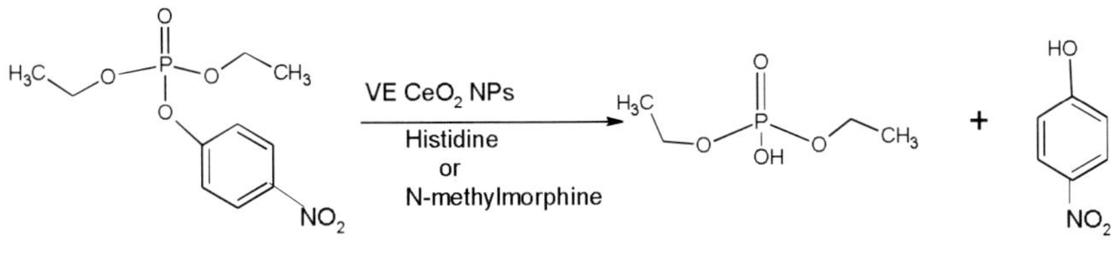

Scheme 1. Degradation of paraoxon via nanoceria.

Vernekar et al. [26] reported the degradation of nerve agents, i.e., paraoxon by the help of vacancy-engineered nanoceria (VE CeO_2 NPs) via hydrolysis of paraoxon leading to non toxicbyproducts. Nanoceria mimics like phosphotriestrase catalyst. The VE CeO_2 NPs possess cerium ions in dual oxidation states which efficiently catalyze the hydrolysis of nerve agents. N-methylmorpholine (NMM) as a general base has been employed for the enhanced hydrolytic activity at 25ºC and 45ºC significantly, which is probably due to the activation of metal-bound water molecules by polarization. CeO_2 nanoparticles synthesized by precipitation method without the using H_2O_2 showed very poor activity. These observations clearly imply that capturing the substrate and activating them at the active sites requires metal ions with dual oxidation states and the catalytic hotspots as an essential inbuilt. An optimum level of mixed valency of Ce at the catalytic sites is required for an efficient degradation process. The presence of metal ions in two different oxidation states may also help in the elimination of reaction products from the reaction sites.

Also, pesticides (parathion methyl), triphenyl phosphate and nerve agents (soman, VX) has been reported to be degraded by the help of nanoceria. The extremely dangerous nerve agents VX (*O*-ethyl *S*-[2-(diisopropylamino)ethyl] methylphosphonothioate) and soman (GD, *O*-Pinacolylmethylphosphono-fluoridate) has been efficiently and effectively degraded by nanoceria nanozyme [27-30].

Zhang et al. reported the synthesis of Graphene quantum dots @ Fe_3O_4 nanoparticles mimic peroxidase-like activities for the degradation of phenolic compounds from aqueous solution [31]. Fe_3O_4/CeO_2 nanocomposites for 4-Chlorophenol degeradation [32]. CuO nanoparticles showing phenol degradation [33]. Fe_3O_4MNP/MWNT nanocomposites capably catalyse the oxidative degradation of phenols to insoluble polyaromatic products which can be easily removed from aqueous solutions [34]. All showing peroxidase like activity. Au@Cu_2O core@shell nanocrystals for the degradation of methyl orange [35] Wang et al. prepared vanadate quantum dots-interspersed g-C_3N_4 nanocomposites for efficient inactivation of *Salmonella* by harvesting visible light [36].

Wang et al. [37] reported the degradation of phenolic pollutant by the help of CH-Cu nanozyme with laccase-like activity. CH-Cu nanozymes were synthesized through a hydrothermal method using Cys-His dipeptide and $CuCl_2$ as precursors. The catalytic activity of CH-Cu and laccase for different substrates, aqueous solutions of phenol, *p*-chlorophenol, 2, 6-dimethoxy-phenol, hydroquinone, *o*-nitrophenol and *o*-aminophenol were also determined.

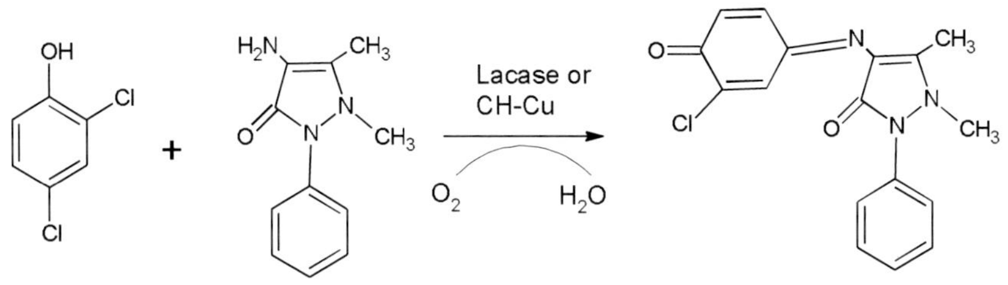

Scheme 2. Degradation of 2,4-dichlorophenol via CH-Cu nanozyme.

2,4-DP and hydroquinone were chosen as the model substrates of the chlorophenols and bisphenols, respectively. The HPLC chromatograms show the degradation of 2,4-DP by the CH-Cu nanozymes with higher degradation efficiency compared with laccase. A rapid, convenient and precise method for the determination of epinephrine concentration has also been established for CH-Cu nanozymes.

Degradation of antibiotics, i.e., Streptomycin, by aptamer modified gold nanoparticle sensor reported by Zhao et al. [38]. Also, activity can be shown for tetracycline, oxytetracycline, carbamazepine, penicillin, amgoxicillin and diclofenac. Detection of kanamycin antibiotic by ssDNAaptamers @ gold nanoparticles reported by Sharma et al. [39].

Geng et al. [40] demonstrated efficient degradation of dye, i.e., methyl orange by copper nanozyme (CNZ) mimicking peroxidase activity. It has been reported that copper nanoparticle exhibit high photocatalytic ability and also show degradation of several dyes, including methyl red, Congo red, and methylene blue [41]. 93% of the degradation rate could be obtained in less than 10 min under the optimum conditions of pH, 3.0; T, 60°C; H_2O_2 concentration, 200 mM; dosage of CNZ, 8 mg. SEM-EDX, FTIR confirmed the morphology and synthesis of CNZ. The kinetic parameters of CNZ were detected by the Michaelis-Menten model. The decolorization rate increased gradually with the increase of catalyst. Under optimum conditions CNZ shows higher activity in comparison to other catalyst like HRP.

Zhu et al. [42] reported a broad classification of nanozymes biosensing Pesticides and Cholinesterases. Nanozymes explored for pesticide analysis mimic as oxidoreductases and hydrolases. Oxidoreductases commonly used in pesticide detection include oxidase and peroxidase such as Fe_3O_4 magnetic nanoparticles with horseradish peroxidase (HRP)-like activity catalyzes the oxidation of 3,3,5,5-tetramethylbenzidine (TMB), 3,3-diaminobenzidine (DAB), o-phenylenediamine (OPD) and 2,2-azobis(3-ethylbenzothiazolin-6-

sulfonic acid) (ABTS) [43, 44]. While phosphatase mimics can be employed as a class of hydrolases simply hydrolyzing Ops, i.e., degrading pesticides [45].

Zhu et al. reported the degradation of five pesticides, i.e., lactofen, fluroxypyr-meptyl, bensulfuron-methyl, fomesafen and diafenthiuron by employing three graphene materials including nitrogen-doped graphene (NG), nitrogen and sulfur co-doped graphene (NSG), and graphene oxide (GO) with peroxidase-like activities. Sensor arrays were successfully used to discriminate each pesticide with different concentrations [46].

Chen et al. [47] reported colorimetric detection methods for organophosphorus pesticides and oxytetracycline by using Cerium oxide nanoparticles mimicking as oxidase-like nanozymes. CeO_2 NPs using the MOFs template strategy were synthesized. Zeoliticimidazolate framework-8 (ZIF-8) NPs was used as the template. The Ce^{3+}/Ce^{4+} ions were hydrolyzed into $Ce(OH)_3/Ce(OH)_4$ under alkaline conditions, which resulted from the hydrolysis of 2-methylimidazole on the surface of the ZIF-8 template. The oxidase-like activity of CeO_2 NPs was also tested over the catalytic oxidation of TMB and ABTS, turning colorless solution of TMB or ABTS to deep blue and green color respectively.

Wu et al. reported the synthesis of Single-atom nanozymes (SAzymes) @high-concentration Cu sites on carbon nanosheets mimicking peroxidase like activity. Three-enzyme-based cascade reaction system, i.e., Cu−N−C SAzymes, natural acetylcholinesterase and choline oxidase enzyme (ACC system) was assembled for the colorimetric detection of acetylcholine and organophosphorus pesticides, i.e., paraoxon-ethyl. AFM results show that the thickness of Cu−N−C is about 1.1 nm. Also TEM and energy dispersive X-ray spectroscopy (EDS) confirmed the synthesis of Cu−N−C nanosheets. The Cu content of Cu−N−C SAzymes was found to be 5.1 wt % by ICP-OES which is quite appreciable. At pH 3.0 an temperature 37°C, nanozyme shows highest activity towards the substrate. The minor K_M (Michaelis-Menton constant) values of Cu−N−C SAzymes indicate stronger affinity toward the substrate. Nanozymes shows recyclability even after 4 runs of activity.

$$\text{ACh} \longrightarrow \text{Choline} \longrightarrow \text{betaine} + H_2O_2$$

For the detection and degradation by ACC complex nanozyme Acetylcholine (ACh) was converted into betaine and peroxide. The sensitive detection of ACh at trace ultralow amounts has been achieved [48].

Conclusion

Due to increased usage and contamination of various organic pollutants it has become an ultimate need for their detection at ultra-low concentrations an also degradation instantly. The alarming concentrations of organic pollutant directly affect our nervous system and shows genetic problems. Harmful and illegal usage has caused many unwanted deteriorating effects on our environment. Green chemistry urges for the use of green methods and compounds which should be environment friendly and recyclable. Therefore, using nanozymes as sensors and degrading tool is utmost present need. Most of the organic pollutants and their detection as well as degradation by various nanozymes have been enlisted. Each nanozymes mimics different type of mechanism for the degradation of OCs.

References

[1] Manea F., Houillon F. B., Pasquato L., and Scrimin P. (2004), *Angew. Chem. Int.* Ed., 43, 6165-6169.

[2] Wei H., Wang, E. (2013) Nanomaterials with enzyme-like characteristics (nanozymes): next-generation artificial enzymes. *Chem. Soc. Rev.*, 42, 6060-6093.

[3] Wu J., Wang X., Wang Q., Lou Z., Li S., Zhu Y., Qin L., Wei H. (2019) Nanomaterials with enzyme-like characteristics (nanozymes): next-generation artificial enzymes (II). *Chem. Soc. Rev.,* 48, 1004-1076.

[4] Harrad S. (2010) *Persistent Organic Pollutants*; Wiley: Chichester, UK.

[5] Feng Y. B., Hong L., Liu A. L., Chen W. D., Li G. W., Chen W., Xia X. H. (2015) High-efficiency catalytic degradation of phenol based on the peroxidase-like activity of cupric oxide nanoparticles. *Int. J. Environ. Sci. Technol.* 12, 653-660.

[6] He J., Liang M. (2020) Nanozymes for environmental monitoring and treatment. In: *Nanozymology*; Yan, X., Ed.; Springer: Berlin, Germany, 527-543.

[7] Cheng R., Li G. Q., Cheng C., Shi L., Zheng X., Ma Z. (2015) Catalytic oxidation of 4-chlorophenol with magnetic Fe_3O_4 nanoparticles: Mechanisms and particle transformation. *RSC Adv.* 5, 66927-66933.

[8] Shin H. Y., Park T. J. and Kim M. I. (2015) *Journal of Nanomaterials*, 2015, 1-11.

[9] Li J. N., Liu W. Q., Wu X. C., Gao X. F. (2015) Mechanism of pH-switchable peroxidase and catalase-like activities of gold, silver, platinum and palladium. *Biomaterials*, 48, 37-44.

[10] Fan K., Xi J., Fan L., Wang P., Zhu C., Tang Y., Xu X., Liang M., Jiang B., Yan X. (2018) *In vivo* guiding nitrogen-doped carbon nanozyme for tumor catalytic therapy. *Nat. Commun.*, 9, 1-11.

[11] Wong E. L. S., Vuong K. Q., and Chow E. (2021) Nanozymes for Environmental Pollutant Monitoring and Remediation, *Sensors* 21, 408.

[12] Borthakur P., Boruah P. K., Dasa P., and Das M. R. (2021) CuS nanoparticles decorated MoS_2 sheets as an efficient nanozyme for selective detection and photocatalytic degradation of hydroquinone in water, *New J. Chem.*

[13] Cai S., Han Q., Qi C., Wang X., Wang T., Jia X., Yang R., and Wang C. (2017) *Chin. J. Chem.*, 35, 605-612.

[14] Zhao L., Xu H., Jiang B. and Huang Y. (2017) *Part. Part. Syst. Charact.*, 34, 1600323.

[15] Geng Q., Guo Q., Cao C. and Wang L. (2008) *Ind. Eng. Chem. Res.*, 47, 2561-2568.

[16] Emara M. M., El-Moselhy M. M. and Farahat N. S. (2010) *Des. Water Treat.*, 19, 232-240.

[17] Wang X., Zhao M., Song Y., Liu Q., Zhang Y., Zhuang Y., and Chen S. (2019) *Sens. Actuators, B*, 283, 130-137.

[18] Ding N., Yan N., Ren C., and Chen X. (2010) Colorimetric determination of melamine in dairy products by Fe_3O_4 Magnetic nanoparticles-H_2O_2-ABTS detection system, *Analytical Chemistry*, 82(13) 5897-5899.

[19] Kim M. I., Ye Y., Won B. Y., Shin S., Lee J., and Park H. G. (2011) A highly efficient electrochemical biosensing platform by employing conductive nanocomposite entrapping magnetic nanoparticles and oxidase in mesoporous carbon foam, *Advanced Functional Materials*, 21(15) 2868-2875.

[20] Zhang J., Zhuang J., Gao L. (2008) Decomposing phenol by the hidden talent of ferromagnetic nanoparticles, *Chemosphere*, 73(9) 1524-1528.

[21] Jiang J. Z., Zou J., Zhu L. H., Huang L., Jiang H., and Zhang Y. (2011) Degradation of methylene blue with H_2O_2 activated by peroxidase-like Fe_3O_4 magnetic nanoparticles, *Journal of Nanoscience and Nanotechnology*, 11(6), 4793-4799.

[22] Zhang J., Zhuang J., Gao L. (2008) Decomposing phenol by the hidden talent of ferromagnetic nanoparticles, *Chemosphere*, 73(9) 1524-1528.

[23] Jiang J. Z., Zou J., Zhu L. H., Huang L., Jiang H., and Zhang Y. (2011) Degradation of methylene blue with H_2O_2 activated by peroxidase-like Fe_3O_4 magnetic nanoparticles, *Journal of Nanoscience and Nanotechnology*, 11 (6) 4793-4799.

[24] Gao L., Giglio K. M., Nelson J. L., Sondermann H., and Travis A. J. (2014) Ferromagnetic nanoparticles with peroxidase-like activity enhance the cleavage of biological macromolecules for biofilm elimination, *Nanoscale*, 6(5) 2588-2593.

[25] Vickery K., Pajkos A., and Cossart Y. (2004) Removal of biofilm from endoscopes: evaluation of detergent efficiency, *American Journal of Infection Control*, 32(3) 170-176.

[26] Vernekar A. A., Das T., and Mugesh G. (2016) Vacancy-Engineered Nanoceria: Enzyme Mimetic Hotspots for the Degradation of Nerve Agents, *Angew. Chem. Int. Ed.*, 55, 1412-1416.

[27] Janoš P., Ederer J., Došek M. (2014) Some Environmentally Relevant Reactions of Cerium Oxide, *Nova Biotechnologica et Chimica* 13-2, 148-161.

[28] Janos P., Kuran P., Pilarova V., Trogl J., Stastny M., Pelant O., Henych J., Bakardjieva S., Zivotsky O., Kormunda M., Mazanec K. and Skoumal M. (2015) *Chem. Eng. J.*, 262, 747-755.

[29] Janos P., Kuran P., Kormunda M., Stengl V., Grygar T. M., Dosek M., Stastny M., Ederer J., Pilarova V. and Vrtoch L. (2014) *J. Rare Earths*, 32, 360-370.

[30] Janos P., Henych J., Pelant O., Pilarova V., Vrtoch L., Kormunda M., Mazanec K. and Stengl V. (2016) *J. Hazard. Mater.*, 304, 259-268.

[31] Wu X. C., Zhang Y., Han, T., Wu H. X., Guo S. W., Zhang J. Y. (2014) Composite of graphene quantum dots and Fe_3O_4 nanoparticles: Peroxidase activity and application in phenolic compound removal. *RSC Adv.*, 4, 3299-3305.

[32] Xu L., Wang J. (2012) Magnetic nanoscaled Fe_3O_4/CeO_2 composite as an efficient Fenton-like heterogeneous catalyst for degradation of 4-chlorophenol. *Environ. Sci. Technol.*, 46, 10145-10153.

[33] Feng Y. B., Hong L., Liu A. L., Chen W. D., Li G. W., Chen W., Xia X. H. (2015) High-efficiency catalytic degradation of phenol based on the peroxidase-like activity of cupric oxide nanoparticles. *Int. J. Environ. Sci. Technol.*, 12, 653-660.

[34] Chen T. M., Xiao J., Wang G. W. (2016) Cubic boron nitride with an intrinsic peroxidase-like activity. *RSC Adv.*, 6, 70124-70132.

[35] Kuo M. Y., Hsiao C. F., Chiu Y. H., Lai T. H., Fang M. J., Wu J. Y., Chen J. W., Wu C. L., Wei K. H., Lin H. C., Hsu Y. J. (2019) Au@Cu_2O core@shell nanocrystals as dual-functional catalysts for sustainable environmental applications, *Appl. Catal. B Environ.* 242 499-506.

[36] Wang R., Kong X., Zhang W., Zhu W., Huang L., Wang J., Zhang X., Liu X., Hu N., Suo Y., Wang J. (2018) Mechanism insight into rapid photocatalytic disinfection of Salmonella based on vanadate QDs-interspersed g-C_3N_4 heterostructures, *Appl. Catal. B Environ.* 225,228-237.

[37] Wang J., Huang R., Qi W., Su R., Binks B. P. and He Z. (2019) Construction of a bioinspired laccase-mimicking nanozyme for the degradation and detection of phenolic pollutants, *Applied Catalysis B: Environmental.*

[38] Zhao J., Wu Y., Tao H., Chen H., Yang W., Qiu, S. (2017) Colorimetric detection of streptomycin in milk based on peroxidase-mimicking catalytic activity of gold nanoparticles. *RSC Adv.*, 7, 38471-38478.

[39] Sharma T. K., Ramanathan R., Weerathunge P., Mohammadtaheri M., Daima H. K., Shukla R., Bansal, V. (2014) Aptamer-mediated 'turn-off/turn-on' nanozyme activity of gold nanoparticles for kanamycin detection. *Chem. Commun.*, 50, 15856-15859.

[40] Geng X., Xie X., Liang Y., Li Z., Yang K., Tao J., Zhang H. and Wang Z. (2021) Facile Fabrication of a Novel Copper Nanozyme for Efficient Dye Degradation, *ACS Omega*, 6, 6284-6291.

[41] Fathima J. B., Pugazhendhi A., Oves M., Venis, (2018) R. Synthesis of eco-friendly copper nanoparticles for augmentation of catalytic degradation of organic dyes. *J. Mol. Liq.*, 260, 1-8.

[42] Zhu H., Liu P., Xu L., Li X., Hu P., Liu B., Pan J., Yang F., Niu, X. Nanozyme-Participated Biosensing of Pesticides and Cholinesterases: A Critical Review. *Biosensors* 2021, 11, 382.

[43] Gao, L. Z. Zhuang J., Nie L., Zhang J. B., Zhang Y., Gu N., Wang T. H., Feng J., Yang D. L., Perrett S., Yan X. (2007) Intrinsic peroxidase-like activity of ferromagnetic nanoparticles. *Nat. Nanotechnol.*, 2, 577-583.

[44] Wei H., Wang E. K. Fe_3O_4 magnetic nanoparticles as peroxidase mimetics and their applications in H_2O_2 and glucose detection. *Anal. Chem.* 2008, 80, 2250-2254.

[45] Liu H. Y., Liu J. W. (2020) Self-imited phosphatase-mimicking CeO_2 nanozymes. *Chem Nano Mat.*, 6, 947-952.

[46] Zhu Y., Wu J., Han L., Wang X., Li W., Guo H. and Wei H. (2020) Nanozyme sensor arrays based on heteroatom doped graphene for detecting pesticides, *Analytical Chemistry*, 1-19.

[47] Chen Z., Wang Y., Mo Y., Long X., Zhao H., Su L., Duan Z., Xiong, Y. (2020), ZIF-8 directed templating synthesis of CeO_2 nanoparticles and its oxidase-like activity for colorimetric detection. *Sens. Actuators B Chem.*, 323, 128625.

[48] Wu Y., Wu J. B., Jiao L., Xu W. Q., Wang H. J., Wei X. Q., Gu W. L., Ren G. X., Zhang N., Zhang, Q. H. (2020) Cascade reaction system integrating single-atom nanozymes with abundant Cu sites for enhanced biosensing. *Anal. Chem.*, 92, 3373-3379.

Chapter 4

Nanozyme-Based Strategies for Environmental Pathogen Detection

C. Pushpalatha[1,*], S. V. Sowmya[2], Dominic Augustine[2], Arshiya Shakir[1], Vysmaya Dhareshwara[1], Amulya Rai[3], Vivek Padmanabhan[4] and Ishitha Singh[2]

[1] Department of Pedodontics and Preventive Dentistry, Faculty of Dental Sciences, M.S. Ramaiah University of Applied Sciences, Bengaluru, India
[2] Department of Oral Pathology and Microbiology, Faculty of Dental Sciences, M. S. Ramaiah University of Applied Sciences, Bengaluru, India
[3] Consulting Endodontist, Ramaiah Memorial Hospital, Bengaluru, India
[4] RAK College of Dental Sciences, RAK Medical and Health Sciences University, Ras al-Khaimah, UAE

Abstract

Nanozymes have exploded in popularity over the years as a group of enzymes that may demonstrate both catalytic capabilities and nanoscale material characteristics. Nanozymes can be mass manufactured at a lower cost than natural enzymes with higher stability. Nanozymes are being more widely used in the field of environmental pathogen analysis as a novel application. Peroxidase, oxidase, superoxide dismutase, and catalase are examples of natural enzymes that work similar to these nanozymes.

For clinical care, homeland security, food safety, and environmental management, it is critical to identify infections quickly, sensitively, and selectively. The development of quick, precise, and cost-effective

---

* Corresponding Author's Email: drpushpalatha29@gmail.com.

In: Emerging Environmental Applications of Nanozymes
Editors: Seema Nara and Smriti Singh
ISBN: 979-8-88697-552-9

technologies for detecting harmful bacteria is critical. In the realm of inorganic enzyme mimics, significant progress has recently been made with the recognition of nanozymes. Because of its features like variable catalytic activity, increased stability, reasonable cost, and ease of synthesis, nanozymes might induce fungal death.

This chapter highlights the various detection methods for bacteria, fungi, protozoa and cancer cells using nanozymes. The applications of nanozymes in antifungal activity, the mechanisms involved and future scope are also discussed. The recent approaches and future implications of nsnozymes-based environmental pathogen detection has been discussed.

Keywords: pathogen detection, aptasensors, immunosensors, nanozymes

1. Introduction

Rapid diagnosis of any disease has been studied extensively in the medical world for many years. Early detection leads to a slew of positive outcomes, including therapeutic health-care decisions and reduced disease severity and morbidity. As a result, various advanced biosensing technologies have been developed for the detection of illness specific biomarkers in order to enable early diagnosis. ELISA, RT-PCR, immunosensors, DNA microarrays, and other diagnostic technologies are used to identify infections, biomolecules, bio-threat chemicals, and poisons (Rodovalho et al., 2015).

The enzyme linked immunosorbent test is one of the most extensively used procedures for recognizing and measuring antigens and peptides (ELISA). Horseradish Peroxidase (HRP) is a natural enzyme that is often used in ELISA. It has a few drawbacks, including proteolytic degradation, inactivation owing to certain preservatives, and limited enzymatic activity in terms of pH and temperature. Furthermore, it is unaffected by antigens and biomolecules at ultra-low concentrations (initial period of disease) (Das et al., 2021).

Nanozymes, or artificial enzymes based on nanomaterials, have aroused researchers' interest due to their broad range of functions and ability to overcome natural enzyme restrictions. Nanozymes bind to certain antigens and generate chromogenic complexes that reveal whether the antigen is present or not. Colorimetric antigen identification provides a quick result for visual analysis with the naked eye as well as measurement of the antigen's UV-visible component. The naked eye's identification could be utilized as a primary

nanozymes catalysing the oxidation of chromogenic substrates, which then change color in response to the concentration of targets.

For example, a 'Dual Lateral Flow Immunoassay' (LFIA) is used to simultaneously detect *E. coli O157:H7* and *Salmonella enteritidis* (Cheng et al., 2017). Target identification and signal amplification are two functionalities of the nanozyme used in LFIA i.e., antibody modified Pt-Pd NPs. Due to the peroxidase-like activity of Pt-Pd nanoparticles in the presence of H_2O_2 and TMB, the nanoparticle-antibody combination turns blue after target recognition. Smartphone-based technology was used to analyse and record the colorimetric observatories. Table 1 lists some of the additional nanozymes that are used in pathogen-specific colorimetric immunosensors.

Table 1. Nanozymes used in immunosensors specific to pathogens

Sl. No	Nanozyme (used in immunosensor)	Pathogen Detected	Author/Year
1.	AuNPs	Influenza A Virus Avian Influenza Virus A	Moulick A et al., 2017
2.	Fe_3O_4 MNPs	*Enterobacter sakazakii* *Ebola Virus*	Zhang L et al., 2017 Duan D et al., 2015
3.	PBNPs	*Salmonella typhimurium*	Farka Z et al., 2018
4.	PtNPs	Human Immuno-Deficiency Virus	Zhang Y et al., 2020

3.2. Fluorescent Immunoassays and Immunosensors Based on Nanozymes

Nanozyme-based fluorescent immunoassays and immunosensors can provide quantitative indicators such as lifespan, polarization, and intensity. Non-fluorescent substrates can create oxidized products with intense fluorescence when catalytic nanozymes are present. Peroxidase-like noble metal nanozymes, for example, dimerize 3-(4-dihydroxyphenyl) propionic acid (HPPA) to produce light by-products (Mahmudunnabi et al., 2020; Olha Demkiv et al., 2020).

3.3. Chemiluminescent Immunoassays and Immunosensors Based on Nanozymes

Chemiluminescence is a form of optical phenomenon that occurs when chemical reactions occur. Chemiluminescence has a wide range of applications, strong selectivity, a quick reaction time, and improved sensitivity. For chemiluminescent analysis, HRP has traditionally been utilized to catalyze the reaction. Because nanozyme-based chemiluminescent immunoassays are more stable, they are used in biosensing (Niu et al., 2020).

3.4. Electrochemical Immunoassays and Immunosensors Based on Nanozymes

Electrochemical immunoassays measure changes in electrochemical signals such as current, potential difference, resistance, and so on before and after antigen-antibody interactions. In these immunosensors, the nanozyme electrochemically catalyses H_2O_2, resulting in extremely sensitive current signals for quantitative assessment (Ronkainen-Matsuno et al., 2007).

4. Nanozyme-Based Aptasensors for Pathogen Detection

In 1990, two separate groups of researchers Ellington et al., and Tuerk et al., discovered a novel class of nucleic acids known as 'Aptamers.' Single-stranded DNA, RNA, or molecular peptides can form a two-dimensional or three-dimensional structure that reacts with the target. Because of the capacity of aptamers to bind to nanozymes, a variety of colorimetric and electrochemical tests have been developed.

In 2019, Bansal and colleagues created an electrochemical sensor to detect the bacterial disease *Pseudomonas aeruginosa* (PA). They paired a PA-specific aptamer (F23) with the nanozyme activity of AuNPs to achieve great affinity and specificity (Chatterjee et al., 2019). Due to the electrochemical properties of TMB, certain aptamers in combination with AuNPs have been shown to aid in the dual detection of PA via colorimetric and electrochemical analysis (Sun et al., 2018). Use of aptamers and AuNP to identify bacteria has been the subject of extensive investigation.

It has been observed that Ochratoxin A (OTA) is a foodborne mycotoxin found in agricultural products such as coffee, cereal grains, dried fruits and wine. OTA is produced by fungi such as *Aspergillus ochraceus*, *A. carbonarius*, *A. niger* and *Penicillium verrucosum*. As OTA is toxic and carcinogenic, its detection is important. Tian et al., in 2019 described an aptasensor for colorimetric OTA detection based on a nanozyme-based reaction.

5. Pathogen-Nanozyme Interactions Employed in Pathogen Detection

In conventional pathogen detection using immunoassay techniques, antigen-antibody conjugates are formed. Antibodies linked to nanozymes bind to certain antigens. This nanozyme catalyzes the chromogenic process, which produces colorimetric changes that are proportionate to the amount of target antigen present. Nanozymes, unlike normal enzymes, allow these reactions to be amplified, making nanozyme-linked immunosensors extremely sensitive to their specific target antigen (Figure 1) (Niu et al., 2020).

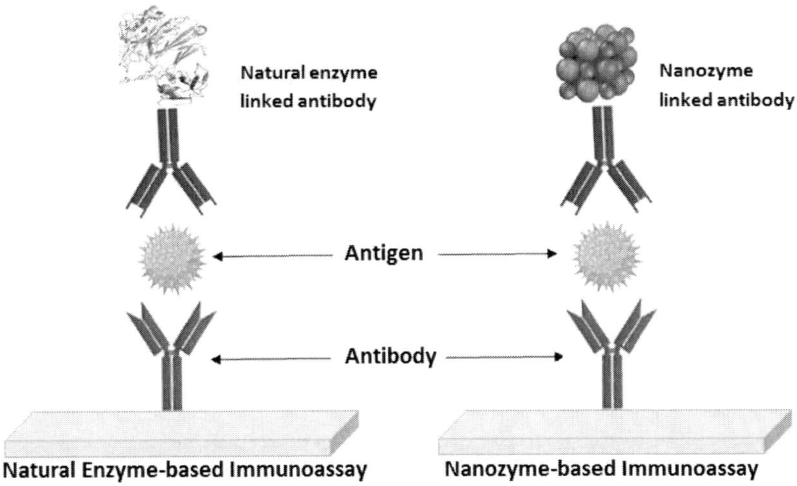

Figure 1. Schematic depiction of both natural enzyme-based immunoassay and nanozyme-based immunoassay.

6. Nanozyme-Based Cancer Cell Detection

Cancer cells do not strictly come under the ambit of pathogens, however, the metastatic process that occurs is included under cancer pathogenesis. The detection and identification of cancer cells can involve use of different methods. Detection using graphene oxide decorated with platinum nanoparticles that are functionalized with folic acid has been reported (Gutiérrez de la Rosa et al., 2022). This method has high specificity and permits a visual colorimetric detection of cancer cells through enhanced nanoconjugation.

The other method that is specific to cancer cell detection is a copper oxide nanozyme supported on a nanocomposite consisting of graphene oxide decorated with gold nanoparticles. This method involves amplification of the electrochemical signal using cyclic voltammetry for the detection of MCF-7 cells (Songca, 2022) as shown in Figure 2.

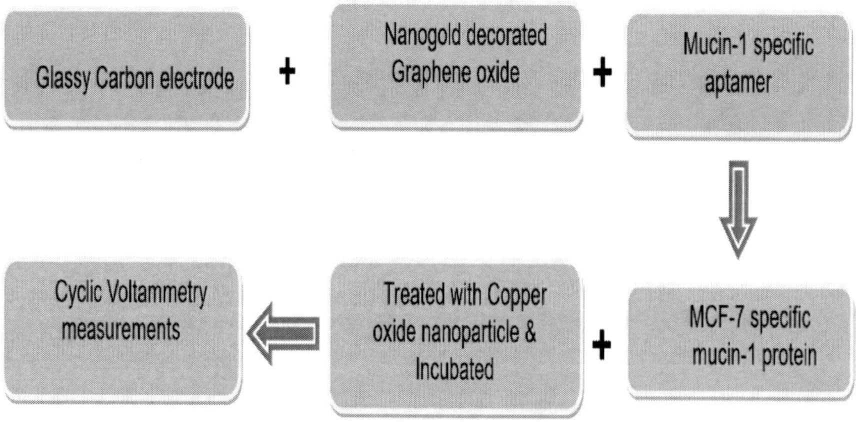

Figure 2. Cyclic voltametry-based cancer cell detection.

7. Nanozyme Platform for Bacterial Identification

A sandwich type immunoassay technique has been reported to detect *Escherichia coli* by mimicking the peroxidase enzyme activity (Zhou et al., 2022). A study done by Wang et al., 2018, have made use of hybrid nanoflowers of hemin and concanavalin A for secondary antibody surrogate detection. This method has shown high selectivity to *E. coli* than other bacteria

(Wang et al., 2018). It makes use of magnetic beads and nanozyme peroxidase conversion of colorless divalent 2,2′-azino-bis 3-ethylbenzthia-zoline-6-sulfonic acid (ABTS) from colorless to green that facilitates detection (Songca, 2022) as depicted in Figure 3.

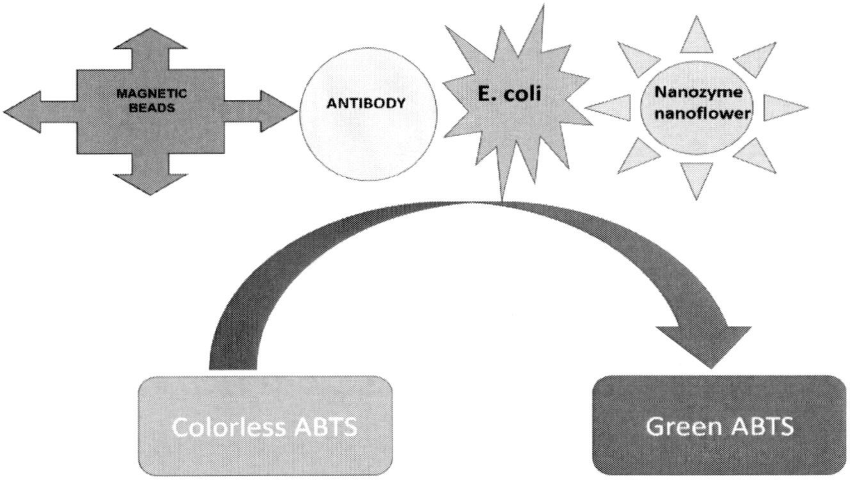

Figure 3. Colorimetric method of bacterial pathogen detection.

8. Nanozyme-Based Parasite Ova Detection

The techniques involved in the detection of parasite ova indirectly in water are graphic microscopic enumeration, PCR, flow cytometry and recombinase polymerase amplification mechanism in a lateral flow strip. However, currently nanozymology has been employed to detect parasites using nanogold in chronoamperometric procedure. For specific detection of *Leishmania* parasites, casein conjugated gold nanoparticles are used. Nanogold-based ELISA has been particularly helpful for the identification of *Trichinella spiralis* (Songca, 2022).

9. Fungi Responses to Enzyme-Mimicking Nanozymes

Antifungal drugs act on fungal infections, but have serious side effects such as damage to kidney and liver. Antibiotic usage has resulted in the emergence of single- and multi-drug-resistant fungus. Natural enzymes originating from plants or bacteria have the power to break down and potentially destroy fungal cell walls. They do, however, have substantial limitations that limit their use, such as high cost, low stability, and difficulty in mass production (Wang et al., 2019). To overcome these challenges, natural enzymes have been replaced by artificial enzymes. In 2007, researchers found Fe_3O_4 nanoparticles with intrinsic peroxidase (POD) activity. Yan's lab later created the notion of "nanozymes," which are artificial enzyme mimics with nanomaterial properties and catalytic function.

9.1. Nanozymes with Antifungal POD Activity

Nanozymes with antifungal activity exhibit a variety of enzymatic activities, with POD-like activity being the most common. Fe_3O_4 nanozymes combined with H_2O_2 can kill fungus depending on the activity of POD-like enzymes. When Drozd et al. in 2016 compared the effects of H_2O_2 and nanozyme alone, they found that combining the two might increase the inhibition rate to 75.70 percent. Snowball-like hybrid nanostructures (NSBs) produced from *Viburnum opulus* and Cu^{2+} ions were also shown to have catalytic activity. The lowest inhibitory concentration of NSB against fungus (Qingzhi et al., 2021) could be lower than the 10 g/ml minimum inhibitory concentrations of NSBs against *E. coli* and *Staphylococcus aureus*. Nanozymes that behave similarly to POD could be used to destroy fungi, gram-positive and negative bacteria.

9.2. Antifungal Activity by Nanozymes with Multienzyme-Like Activity

Immune cells kill fungus by producing reactive oxygen species through the catalysis of NADPH oxidase and myeloperoxidase (MPO). Nanozymes possessing multienzyme-like activity cause fungal death by generating reactive oxygen species (ROS). The nanozyme having Ce-metal-organic framework (Ce-MOF) suppressed 93.3–99.3 percent of fungal growth and produced hyphae and conidiophore deformation by disrupting the fungal cell wall membrane and causing oxidative stress in the fungal cell (Wei et al., 2021). Cu–Fe NP causes fungal inactivation in a concentration and time dependent manner. Cu–Fe NPs can penetrate fungal cells, where the liberated iron and copper combine with ROS (O_2) to create H_2O_2, which is catalyzed by SOD. Copper and iron converted H_2O_2 to ROS via Fenton and Haber–Weiss chemistry, resulting in fungal death. These findings imply that the nanozymes with SOD, OXD, and POD-like activities can catalyze a cascade reaction that converts O_2 into a dangerous radical with antifungal activity (Qingzhi et al., 2021)

9.3. Antifungal Nanozymes with Additional Enzyme-Like Functions

Fungal cell walls have been reported to be attacked by nanozymes with enzymatic activity. Li et al., in 2015 were able to disintegrate the cell walls of yeast cells using Fe_3O_4MNPs with zymolyase-like lytic activity. Demkiv et al. (2021) developed nano-emulsions that functioned similarly to glutathione POD and discovered that they were efficient against fungus. Deformation and wrinkling caused fungal cells to lose their shape. The nano-emulsion was then able to reduce biofilm formation by 86.2 percent while also controlling adhesion rates at 35.7–57.4%.

10. Mechanisms of Nanozyme Antifungal Effect

Nanozymes are nanomaterials with built-in enzyme activity that catalyse natural enzyme substrates and follow the same reaction kinetics as their natural counterparts. The enzyme-like activity of a nanozyme stems from its distinct nanostructure, which eliminates the need for additional catalytic groups. More enzyme-like nanomaterials have recently been identified, which may be classified into four categories: hydrolase, oxidoreductase, lyase, and isomerase, with oxidoreductase resembling activity being significant, including CAT, SOD, OXD, and POD (Huang et al., 2019). Nanozymes, like natural enzyme's antifungal processes, heavily catalyse the matching substrate to create ROS to fight fungal invasion. Although ROS play an important role in human biology, excessive amounts can cause normal cells to perish or apoptose (Fang et al., 2011; Duan et al., 2021).

10.1. Functions of a Nanozyme to Manage ROS

Superoxide anion (O_2^-) hydroxyl radical (OH), H_2O_2, and singlet oxygen are examples of intermediate products. Nanozymes have been produced with a variety of enzyme activities to control ROS levels in order to protect normal cells or kill malignant cells (Zhang et al., 2013; Zhao et al., 2020). Nanomaterials with POD and OXD activities produce ROS in the presence of H_2O_2 or O_2, which are scavenged by SOD, which transforms superoxide into H_2O_2 and O_2 (Wang et al., 2018). It is critical to understand that these effects are pH-dependent. Because increased ROS is required to kill fungal cells, yet too much ROS can harm host cells, when employing nanozymes for antifungal therapy, the local microenvironment must be considered. Furthermore, a single nanozyme can perform a variety of enzyme-like functions, favoring ROS formation through cascade events involving the consumption of a specific substrate. For example, nanozymes with bifunctional glucose oxidase and POD-like activity may produce H_2O_2 from glucose oxidation and then convert H_2O_2 to OH radicals, effectively "killing two birds with one stone". Nanozymes that consume glutathione (GSH) may indirectly increase ROS by affecting the redox balance, which is beneficial for antifungal therapy (Meng et al., 2021). Physical signals, such as chemical activators/inhibitors or the photothermal effect induced by infrared light, can also affect nanozyme activity (Gao et al., 2017).

10.2. Non-ROS Pathway-Based Antifungal Nanozyme

Many nanomaterials with high enzyme-like activity catalyse H_2O_2 to produce damaging radicals to achieve a sterilizing effect. H_2O_2 is a potent oxidant that has been shown to kill pathogens such as *E. coli, S. aureus,* and *Candida albicans.* Fe_3O_4 nanozymes combined with H_2O_2 can kill fungus by creating reactive oxygen species (ROS). The low combination probability of nanozymes with fungus, as well as the large distance between fungi and the produced ROS, limits the antifungal activity of nanozymes (Zhang et al., 2013). Baldemir et al., 2017, discovered that nano-flowers with POD-like activity destroyed cell membranes of fungi by electrostatic contact with OH. TiO_2 NP co-doped with nitrogen and fluorine binds to the fungus surface and kills it by interacting with chitin or glucan in fungal cell walls to produce reactive oxygen species (ROS) (Mukherjee et al., 2020). These OH may cause cell wall damage and allow intracellular contents to seep out by rupturing glycosidic linkages in chitin or glucan. By gradually enhancing POD activity under visible light irradiation, nitrogen–iodine–doped carbon dots with POD-like activity may improve photocatalytic inactivation of *Candida albicans* that is reliant on ROS (OH) (Li et al., 2014). The Fe_3O_4MoS2-Ag nanozyme catches fungal cells by assaulting its cell membrane and damaging DNA with its enzymatic activity and photothermal properties, thanks to its rough surface.

10.3. ROS Pathway-Based Antifungal Nanozyme

Nanozymes demonstrate both ROS-dependent and ROS-independent fungal killing strategies. Xiao et al. used this co-precipitation approach to make Fe_3O_4 MNPs with lytic activity similar to that of zymolyase on yeast cell walls. Hypohalous acids formed from halides, H_2O_2, or OH are thought to be antifungal, and particular transition metal oxide NPs are thought to be halogenating enzyme mimics. However, the specific mechanisms of some nanozymes that work in a ROS-independent manner are unknown. Nano-emulsions with significant antifungal activity, for example, target the components of fungal cells directly or indirectly using products formed by glutathione peroxidase (GSH-Px) catalysis (Qingzhi et al., 2021).

10.4. Benefits and Limitations of Antifungal Nanozymes

Nanozymes rely on catalytic defense against fungal infections as well, but they are easier to make, less expensive, and more stable (Lin et al., 2019). These properties assist natural enzymes in resolving a multitude of difficulties. In a recent study, Fe_3O_4 MNPs with zymolyase-like activity were found to disrupt the cell wall of *Saccharomyces cerevisiae*. Several enzymes can destroy cell membranes and the integrity of fungal cells by creating excessive amounts of reactive oxygen species (ROS), depending on their inherent enzymatic activity. Natural enzymes, on the other hand, are unable to regulate the amounts of ROS. When the formation of ROS is more than the antioxidant capability of cellular antioxidants in biological systems, normal cells are killed. Nanozymes, unlike natural enzymes, can modify pH, temperature, shape, and size to control the generation of ROS. A nanozyme is a one-of-a-kind artificial enzyme that generates reactive oxygen species to destroy cell membrane integrity and remove biofilm matrix components. The fungus wraps its pseudopod around the immune cells, starting the process of fungus destruction by immune cells. In this mechanism, rough surfaces are stickier than smooth surfaces. The rough surface of Fe_3O_4 MoS2-Ag, inspired by innate immune cells, makes it easier to catch fungus (Wei et al., 2021). Nanoflowers, on the other hand, employ electrostatic action to minimize the distance between the fungus and the particles, and then use the Fenton reaction to kill stubborn fungi.

To avoid immune cells in general, fungi use pathogenicity, such as hyphae and invasion enzymes (Ildizet al., 2017). Nanozymes can kill fungus by stopping them from developing hyphae and thereby escape. Nanozymes can destroy *Aspergillus niger, Candida albicans, E. coli, Aspergillus flavus,* and *Staphylococcus aureus*, among other fungus and bacteria (Abdelhamid et al., 2020). The fungus is less likely to create drug-resistant races due to the exact components of nanozymes. When compared to conventional antifungal medicines such as antibiotics and antimicrobial peptides, nanozymes exhibit a strong antifungal effect with no drug resistance.

11. Recent Approaches and Future Implications of Environmental Pathogen Detection Using Nanozymes

Environmental pathogens are the main focus in the recent times due to their impact on causation of many infections in the form of epidemics like cholera and dysentery and the global COVID-19 pandemic. Detection of environmental virulent pathogens before they actually cause infection is the current challenge and will be the future consideration. Many novel diagnostic tests have been developed to replace traditional methods. Nanofluidic real-time PCR system is one such recent method for qualitative and quantitative molecular recognition of 17 human and 2 food borne viruses and has shown better results compared to conventional PCR (Coudray et al., 2016).

An assay has been reported by Asati et al. (2009) for the detection of tumor cells with polyacrylic acid coated nanoceria. Folic acid conjugated nanoparticles bind to folate receptors on cancer cells thereby helping in their identification (Asati et al., 2009).

Detection of fungi which was traditionally performed using culture method has currently been replaced by PCR. A study by Wagner et al., has demonstrated that PCR methods are faster and reliable as compared to the routine culture method. Recently biosensors have the advantage of being highly sensitive, durable, reliable, cost effective and rapid in detection (Wagner et al., 2018).

Colorimetric paper-based biosensors are the recent strategies that comprise of three-variants such as dipstick or spot-test assay, lateral flow assay and paper-based microfluidic analytical devices. Nanozymes can be added to enhance the performance of biosensor devices for pathogen detection. Microfluidic biosensors are gaining a lot of popularity due to the involvement of computerized technology and high precision. Recent incorporation of nanozymology into their design along with analytical strategies for pathogen detection in environmental, food and clinical samples is of great interest (Songca, 2022).

The future concerns with use of novel nanozyme-based environmental pathogen identification are reduced detection limits, lower detection time, use of quality devices and increased point-of care usage. It has been postulated that nanozyme based pathogen recognition has a major role to play in the control of current and future pandemics. Combination of diagnostic and therapeutic nanozyme based strategies helps in the efficient healthcare improvement and reduces the disease burden (Das et al., 2021).

Conclusion

Enzymatic therapy appears to be a viable method for reducing bacterial antibiotic resistance. Nanozymes are a form of nanomaterial that acts like an enzyme but is less expensive and more stable than natural enzymes. Nanozybiotics which are based on nanozymes have demonstrated significant antibacterial application potential towards resistant bacteria by replicating natural enzyme-like actions. Novel biocompatible nanozybiotics that utilize enzyme-like nanozymes with a variety of antibacterial properties and situations are urgently needed.

The biotoxicity of the nanozyme may be reduced by combining it with other biocompatible components. Nanozymes are a brand-new cross-disciplinary field that is only getting started in terms of application. Furthermore, antifungal activity of natural enzymes can be used to create nanozymes. Various detection techniques such as modified PCR, immunosensors, colorimetric immunoassays and aptasensors have gained popularity in the diagnostic field. We anticipate that as research advances and knowledge is transferred from fundamental to clinical practice, nanozymes will emerge as new effective antifungal medications to treat and prevent fungal infections, thereby improving quality of life of all human beings. Nanozyme advances in the future will usher in a new age in the fight against bacteria, protozoa, fungi and cancer cells. Although current nanozyme research is limited, the future of nanozyme technology appears bright.

References

Abdelhamid, H. N., Mahmoud G. A., Sharmouk W.A cerium-based MOFzyme with multi-enzyme-like activity for the disruption and inhibition of fungal recolonization. *J Mater Chem B*. 2020;8(33):7548–56.

Adeniyi, O., Sicwetsha, S.,Mashazi, P. (2020) Nanomagnet-silica nanoparticles decorated with Au@Pd for enhanced peroxidase- like activity and colorimetric glucose sensing. *ACS Appl. Mater. Interfaces* 12(2), 1973–1987.

Ali, J., Elahi,S. N., Ali,A.,Waseem, H., Abid, R. (2021) Mohamed, MM. Unveiling the Potential Role of Nanozymes in Combating the COVID-19 Outbreak. *Nanomaterials*, 11, 1328.

Asati, A, Santra S, Kaittanis C, Nath S, Perez JM. Oxidase-like activity of polymer-coated cerium oxide nanoparticles. *Angewandte Chemie*. 2009 Mar 16;121(13):2344-8.

Baldemir, Ayşe, N. Buse Köse, NilayIldız, Selenİlgün, Sadi Yusufbeyoğlu, Vedat Yilmaz and Ismail Ocsoy. Synthesis and characterization of green tea (Camellia sinensis (L.) Kuntze) extract and its major components-based nanoflowers: a new strategy to enhance antimicrobial activity. *RSC Adv*. 2017;7(70):44303–8.

CaiS, JiaX., Han Q. S., Yan X., Yang R., Wang C., Porous Pt/Ag nanoparticles with excellent multifunctional enzyme mimic activities and antibacterial effects, *Nano Res*. 10 (2017) 2069–2082.

Chatterjee, B., Das, S. J., Anand, A., Sharma, T. K. (2019).Nanozymes and aptamer-based biosensing. *Materials Science for Energy Technologies*. https://doi.org/10.1016/j.mset.2019.08.007.

Cheng, N., Song, Y., Zeinhom, M. M. A., Chang, Y. C., Sheng, L. N., Li, H. L., Du, D., Li, L., Zhu, M. J., Luo, Y. B., Xu, W. T., Lin, Y. H. (2017). Review—Nanozyme-Based Immunosensors and Immunoassays: Recent Developments and Future Trends. *ACS Applied Materials & Interfaces*, 9, 40671.

Coudray-Meunier C., Fraisse A., Martin-Latil S., Delannoy S., Fach P., Perelle S. A novel high-throughput method for molecular detection of human pathogenic viruses using a nanofluidic real-time PCR system. *PLoS ONE*. 2016;11:e0147832.

Das, B., Franco, J.L., Logan, N., Balasubramanian P., Kim M.I., Cao C. Nanozymes in point-of-care diagnosis: An emerging futuristic approach for biosensing. *Nano-Micro Lett*. 2021;13:193.

Demkiv, O, Stasyuk N, Serkiz R, Gayda G, Nisnevitch M, Gonchar M. Peroxidase-Like Metal-Based Nanozymes: Synthesis, Catalytic Properties, and Analytical Application. *Applied Sciences*. 2021 Jan 15;11(2):777.

Duan, D, Fan, K, Zhang, D, Tan, S, Liang, M, Liu, Y, Zhang, J, Zhang, P, Liu, W, Qiu, X, Kobinger, GP, Gao, GF, Yan, X. Nanozyme-strip for rapid local diagnosis of Ebola. *Biosens Bioelectron.* 2015 Dec 15;74:134-41.

Duan, J., Gao, S., Tu, S., Lenahan, C., Shao, A., Sheng, J. Pathophysiology and therapeutic potential of NADPH oxidases in ischemic stroke-induced oxidative stress. *Oxid Med Cell Longev.* 2021;2021:6631805.

Ellington, A.D., Szostak, J.W. (1990).In vitro selection of RNA molecules that bind specific ligands. *Lett.To Nat.* 346.818–822.doi:10.1016/0021-9797(80)90501-9.

Fang, F. C. Antimicrobial actions of reactive oxygen species.*Bio.* 2011;2(5):e00141-11.

Farka, Z, Čunderlová V, Horáčková V, Pastucha M, Mikušová Z, Hlaváček A, Skládal P. Prussian Blue Nanoparticles as a Catalytic Label in a Sandwich Nanozyme-Linked Immunosorbent Assay. *Anal Chem.* 2018 Feb 6;90(3):2348-2354.

Gao, L., Fan K., Yan X. Iron oxide nanozyme:a multifunctional enzyme mimetic for biomedical applications. *Theranostics.*2017;7(13):3207–27.

Gutiérrez de la Rosa, SY, Muñiz Diaz R, Villalobos Gutiérrez PT, Patakfalvi R, Gutiérrez Coronado Ó. Functionalized Platinum Nanoparticles with Biomedical Applications. *Int J Mol Sci.* 2022 Aug 20;23(16):9404.

Huang, Y, Ren J, Qu X. Nanozymes: classification, catalytic mechanisms, activity regulation, and applications. *ChemRev.* 2019;119(6):4357–412.

Lin, Y. Liu, X. Zhu, X. Chen, J. Liu, Y. Zhou, X. Qin, J. Liu, Bacteria-responsive biomimetic selenium nanosystem for multidrug-resistant bacterial infection, detection and inhibition, *ACS Nano* 13 (2019) 13965–13984.

Mahmudunnabi, R. G.,Farhana, F. Z.,Kashaninejad, N.,Firoz, S. H.,Shim Y.,Shiddiky, M. J. A. (2020). Nanozyme-based electrochemical biosensors for disease biomarker detection. *Analyst*, doi: 10.1039/D0AN00558D.

Meng, Xiangqin, Dandan Li, Lei Chen, Helen He, Qian Wang, Chaoyi Hong, Jiuyang He, Xingfa Gao, Yili Yang, Bing Jiang, Guohui Nie, Xiyun Yan,Lizeng Gao and Kelong Fan. High performance self-cascade pyrite nanozymes for apoptosis ferroptosis synergistic tumor therapy. *ACS Nano.*2021;15(3):5735–51.

Moulick A, Richtera L, Milosavljevic V, Cernei N, Haddad Y, Zitka O, Kopel P, Heger Z, Adam V. Advanced nanotechnologies in avian influenza: Current status and future trends - A review. *Anal Chim Acta.* 2017 Aug 29;983:42-53. doi: 10.1016/j.aca.2017.06.045. Epub 2017 Jul 7.

Mukherjee, K., Acharya K., Biswas A., Jana N. R. TiO_2 nanoparticles Co-doped with nitrogen and fluorine as visible-light-activated antifungal agents. *ACS Appl Nano Mater.* 2020;3(2):2016–25.

Niu, X., Cheng, N.,Ruan, X., Du, D; Lin, Y.,Review—Nanozyme-Based Immunosensors and Immunoassays: Recent Developments and Future Trends 2020.*J. Electrochem. Soc.* 167 037508.

Qingzhi, W, Zou S, Wang Q, Chen L, Yan X, Gao L. Catalytic defense against fungal pathogens using nanozymes. *Nanotechnology Reviews.* 2021 Jan 1;10(1):1277-92.

Rodovalho, VR, Alves L, Castro A, Madurro J, Brito-Madurro A, Santos A. Biosensors applied to diagnosis of infectious diseases–An update. *Austin J Biosens Bioelectron.* 2015;1(3):10-5.

Ronkainen-Matsuno, NJ, Halsall HB, Heineman WR. Electrochemical immunoassays and immunosensors. *Immunoassay and Other Bioanalytical Techniques*. 2007:385-402.

Songca SP. Applications of Nanozymology in the Detection and Identification of Viral, Bacterial and Fungal Pathogens. *Int J Mol Sci*. 2022 Apr 22;23(9):4638.

Sun H., Cai S., Wang C., ChenY., YangR., Recent Progress of Nanozymes in Detection of Pathogenic Microorganisms. *Chem Bio Chem*, 21(18), (2018)2572-2584.

Tian, F., Zhou, J., Jiao, B., He, Y., A nanozyme-based cascade colorimetric aptasensor for amplified detection of ochratoxin A. *Nanoscale*, 11, (2019) 9547–9555. doi:10.1039/c9nr02872b.

Tuerk, C., Gold, L. (1990). Systematic evolution of ligands by exponential enrichment: Chemi- SELEX. *Science* (80).249. 505–510. doi:10.1038/346818a0.

Wagner, K., Springer B., Pires V.P., Keller P.M. Molecular detection of fungal pathogens in clinical specimens by 18S rDNA high-throughput screening in comparison to ITS PCR and culture. *Sci. Rep*. 2018;8:6964.

Wang, KY, Bu SJ, Ju CJ, Li CT, Li ZY, Han Y, Ma CY, Wang CY, Hao Z, Liu WS, Wan JY. Hemin-incorporated nanoflowers as enzyme mimics for colorimetric detection of foodborne pathogenic bacteria. *Bioorganic & Medicinal Chemistry Letters*. 2018 Dec 15;28(23-24):3802-7.

Wang, Y, Branicky R, Noë A, Hekimi S. Superoxide dismutases: Dual roles in controlling ROS damage and regulating ROS signaling. J Cell Biol. 2018 Jun 4;217(6):1915-1928.

Wei, F., Cui X., Wang Z., Dong C., Li J., Han X., Recoverable peroxidase- like Fe_3O_4@MoS_2-Ag nanozyme with enhanced antibacterial ability. *Chem Eng J*. 2021; 408:127240.

Wong, ELS, Vuong KQ, Chow E. Nanozymes for Environmental Pollutant Monitoring and Remediation. *Sensors*. 2021; 21(2):408.

Zhang, D., ZhaoY., GaoY., GaoF., FanY., LiX., DuanZ., WangH., Anti-bacterial and *in vivo* tumor treatment by reactive oxygen species generated by magnetic nanoparticles, *J. Mater. Chem.* B 1 (2013) 5100–5107.

Zhang, J., Liu Y., Li Q., Zhang X., Shang J. K. Antifungal activity and mechanism of palladium-modified nitrogen-doped titanium oxide photocatalyst on agricultural pathogenic fungi *Fusarium graminearum*. *ACS Appl Mater Interfaces*. 2013;5(21): 10953–9.

Zhang, L, Chen Y, Cheng N, Xu Y, Huang K, Luo Y, Wang P, Duan D, Xu W. Ultrasensitive detection of viable Enterobacter sakazakii by a continual cascade nanozyme biosensor. *Analytical Chemistry*. 2017 Oct 3;89(19):10194-200.

Zhang, Y, Jin Y, Cui H, Yan X, Fan K. Nanozyme-based catalytic theranostics. *RSC Advances*. 2020;10(1):10-20.

Zhao, S., Yu X., Qian Y., Chen W., Shen J. Multifunctional magnetic iron oxide nanoparticles: an advanced platform for cancer theranostics. *Theranostics*. 2020;10 (14):6278–309.

Zhou, C, Wang Q, Jiang J, Gao L. Nanozybiotics: Nanozyme-Based Antibacterials against Bacterial Resistance. *Antibiotics*. 2022 Mar 15;11(3):390.

Chapter 5

Dye Degradation and Removal by Nanozymes

Jayeeta Chattopadhyay[1,*], Sushant Kumar[1], Tara Sankar Pathak[2], Prachi Priyanka[3] and Nimmy Srivastava[3]

[1] Chemistry Department,
Amity School of Engineering and Technology,
Amity University Jharkhand, Ranchi, India
[2] Department of Science and Humanities,
Surendra Institute of Engineering and Management, West Bengal, India
[3] Amity Institute of Biotechnology,
Amity University Jharkhand, Ranchi, India

Abstract

In recent days, various engineered nano-structures are mimicking natural enzyme-like activities, whose distinctive physical and chemical properties have allowed exceptional technological advancement.

In this chapter, we will cover the recent progress in the field of nanozymes which work remarkably well in dye degradation. It will discuss comprehensive overview on modification of these special structured nanozymes in respect to size, shape, and morphology to achieve desired activity. This chapter will also include the challenges and prospects of potential synthetic strategies of nanozymes by using metal organic frameworks (MOFs).

Keywords: dye degradation, MOF, nanoparticles, metal oxides

---

* Corresponding Author's Email: jchattopadhyay@rnc.amity.edu.

In: Emerging Environmental Applications of Nanozymes
Editors: Seema Nara and Smriti Singh
ISBN: 979-8-88697-552-9

1. Introduction

Nanozymes are sophisticated nanomaterials with unique physicochemical properties and the capacity to imitate intrinsic physiologically relevant events through precise structural creation. Nanozymes, particularly, mimic real enzymes and have enzyme-like characteristics. Enzymatic catalytic reactions are very effective, happening quickly even under mild conditions, and are also highly selective. For sensing and monitoring applications, great efficiency and selectivity are extremely desirable. Natural enzymes, including proteins, have constraints such as limited thermostability and a small pH window, which cause them to denature and diminish or inhibit their enzymatic activity. Low thermostability creates additional demands on natural enzyme storage, transit, and handling, which can be labor and infrastructure-intensive for users.

Susceptible denaturation complicates the interpretation of sensing and monitoring data, perhaps leading to a false positive or negative result. Nanozymes, in this context, address these limits by providing great structural durability and stability while maintaining optimal catalytic activity. Nanozymes are sophisticated nanomaterials with unique physicochemical properties and the capacity to imitate intrinsic physiologically relevant events through precise structural creation. Nanozymes, in particular, mimic real enzymes and have enzyme-like characteristics. Enzymatic catalytic reactions are very effective, happening quickly even under mild conditions, and are also highly selective. For sensing and monitoring applications, great efficiency and selectivity are extremely desirable. Natural enzymes, including proteins, have constraints such as limited thermostability and a small pH window, which cause them to denature and diminish or inhibit their enzymatic activity. Low thermostability creates additional demands on natural enzyme storage, transit, and handling, which can be labor and infrastructure-intensive for users.

Susceptible denaturation complicates the interpretation of sensing and monitoring data, perhaps leading to a false positive or negative result. Nanozymes, in this context, address these limits by providing great structural durability and stability while maintaining optimal catalytic activity. Nanozymes have potential applications in a variety of sectors, including biomedicine (in vivo diagnostics and therapies) and the environment, due to their unique physicochemical features and enzyme-like capabilities (detection and remediation of inorganic and organic pollutants). Nanozymes are attractive candidates for real-time and/or remote environmental monitoring and cleanup because they have higher structural stability than natural enzymes and have wider physical (e.g. temperature) and chemical (e.g. pH) operational windows. This is especially true considering the outdoors' demanding and unpredictable character (compared to a more physiologically stable in vivo or in vitro environment which has a more defined and narrower operational window).

Various types of nanomaterials have been found to have intrinsic enzyme-like activity in the last few decades (Wei et al., 2013; Wu et al., 2019). Natural enzymes have intrinsic catalytic ability to catalyse a specific chemical change, usually at a single active site (Kazlauskas et al., 2005; Khersonsky et al., 2006). Because nanozymes lack an active site, researchers have proposed a variety of ways to improve the catalytic characteristics of these nanomaterials, allowing them to react with target molecules selectively and effectively. Nanozymes have been categorized into four groups here and represented in Figure 1.

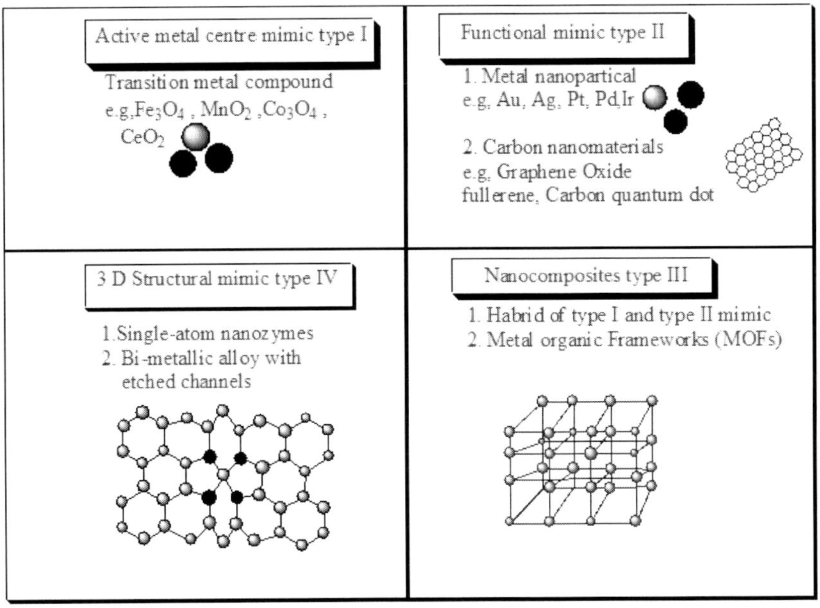

Figure 1. Types of nanozymes based on their mode of natural enzyme-mimicking behaviour.

Our environment has been significantly contaminated as a result of human activities (Huag et al., 2019; Hussain et al., 2020; Wang et al., 2016). Many harmful contaminants are produced during the manufacturing and use of consumer goods, including heavy metals, dyes, and other organic and inorganic compounds. Pesticides, medications, and clothing are among them. Many dangerous substances have been dumped into the environment, particularly aquatic settings (lakes, rivers, and seas), without a second thought due to a lack of regulation and knowledge. These contaminants have wreaked havoc on our ecosystems, harming humans, plants, animals, and microbes alike. They have been proven to have harmful and carcinogenic consequences in people as a result of contamination of drinking water and foods (Wang et al., 2016). The remediation activity to eliminate these contaminants and/or degrade them to less dangerous compounds has received a lot of attention (Hussain et al., 2020; He et al., 2020). Adsorption, membrane filtration, distillation, oxidation, biocatalytic, and photocatalytic degradation are all methods for removing pollutants from aquatic settings (Hussain et al., 2020; He et al., 2020).

The effectiveness of biocatalytic techniques (enzymatic and microbial) for the breakdown of organic pollutants has been studied. These biocatalytic technologies have the benefit of being able to work in moderate and natural conditions. Furthermore, when their job is accomplished, microorganisms and enzymes may degrade into benign chemicals, removing undesired environmental contamination from the treatment agent (Huang et al., 2012 and Huang et al., 2019). Microbes and enzymes, on the other hand, can only work and be effective within specific temperature and pH ranges. The high cost of enzyme production, their lack of recyclability, and the fact that biodegradation can take a long time have all hampered their widespread use in environmental pollution treatment (Hussain et al., 2020; Meng et al., 2020). In order to overcome the limits of enzymatic approaches, researchers are looking at employing nanozymes as pollution remediation agents.

Nanozymes have been demonstrated to have catalytic capabilities, such as peroxidase- and oxidase-like activity, which are used by natural enzymes to degrade contaminants. Nanozymes have the potential to overcome enzyme restrictions in terms of cost, recyclability, reaction rate, and operational windows (pH and temperature) (Huang et al., 2019; He et al., 2020; Sharma et al., 2018; Meng et al., 2020). Nanozymes have been shown to be effective at detecting heavy metals and organic pollutants in the environment in the previous section. Nanozymes could also be utilised to breakdown contaminants in the environment. The applications and future prospects of nanozymes in environmental pollutant remediation, with a focus on the degradation of persistent organic pollutants, will be reviewed in this part. Phenolic chemicals, insecticides, dyes, and organophosphorus compounds are examples of persistent organic pollutants (Harrad et al., 2010).

Pesticides, dyes, and other synthetic substances contain phenolic compounds, particularly chlorophenols (Hussain et al., 2020; Cheng et al., 2015). Because chlorophenol pollutants are very hazardous and resistant to chemical and biological breakdown in the environment, they can cause major environmental problems. The US Environmental Protection Agency (EPA) and other environmental authorities across the world have classed these substances as top priority pollutants. Solvent extraction, physical adsorption, pervaporation, wet air oxidation, ozonolysis, wet peroxide oxidation, electrochemical oxidation, photocatalytic oxidation, supercritical water gasification, electrical discharge degradation, and bio-degradation are some of the current methods for phenol removal (Feng et al., 2014). Nanozyme-mediated phenolic compound degradation is a very promising approach with several advantages over current methods (He et al., 2020).

Table 1. Active metal centre mimic (type I) nanozymes in various environmental pollutants remediation

Nanozyme Type	Removal of Pollutant	Enzyme-like Function	Year	Ref.
Fe_3O_4 MNPs	Phenol	Peroxidase	2008	Zhang et al.
Fe_3O_4 MNPs	Rh B	Peroxidase	2019	Wang et al.
Fe_3O_4 MNPs	Sulfamonomethoxine	Peroxidase	2012	Huang et al.
CuO porous structures	Phenol	Peroxidase	2014	Liu et al.
Fe_3O_4 nanorod bundles	Crystal violet	Peroxidase	2015	Bhuyan et al.
Fe_3O_4 nanorods	Rh B, methylene blue and methyl orange	Peroxidase	2012	Xu et al.
CeO_2 nanoparticles	Rh B, fluorescein, xylene cyanol FF, B	Peroxidase	2016	Zeb et al.

2. Active Metal Centre Mimics (Type I)

Table 1 summarises recent research on type I nanozymes for environmental pollution cleanup. Since Zhang et al., (2008) reported that Fe_3O_4 MNPs have enzymatic-like activity similar to naturally occurring peroxidases, they have been widely used in wastewater treatment for the oxidation of organic substrates as a diagnostic method. Due to their peroxidase-like activity and the ease with which Fe_3O_4 MNPs may be made from inexpensive and abundant precursors, iron oxide nanoparticles, particularly Fe_3O_4 MNPs, have been the most widely explored for the degradation of numerous environmental contaminants. As shown in Table 1, the peroxidase-like activity is the most important enzyme-like catalytic activity marshalled by type I nanozymes in the degradation of phenol chemicals and dyes. Only two occurrences (Zanos et al., 2015; Wang et al., 2013) where MnO_2 nanoparticles and Cu complex demonstrated laccase-like activity are non-peroxidase-like examples. Laccases aid in the oxidation of phenolic substances by converting oxygen to water. Of the metal oxides studied for environmental pollutant degradation, Fe_3O_4 was the most prevalent.

In the absence of H_2O_2, Wang et al., (2013) created a laccase-mimicking nanozyme by coupling Cu^+/$Cu2^+$ to cysteine-histidine dipeptide to generate an inorganic polymer CH-Cu with good catalytic efficiency for the degradation of chlorophenol and bisphenol (Figure 2). CH-Cu was also extremely durable, operating at pH 3–9, temperatures ranging from 20 to 90 °C, and high salinity for more than three weeks. The CH-Cu might be utilised several times.

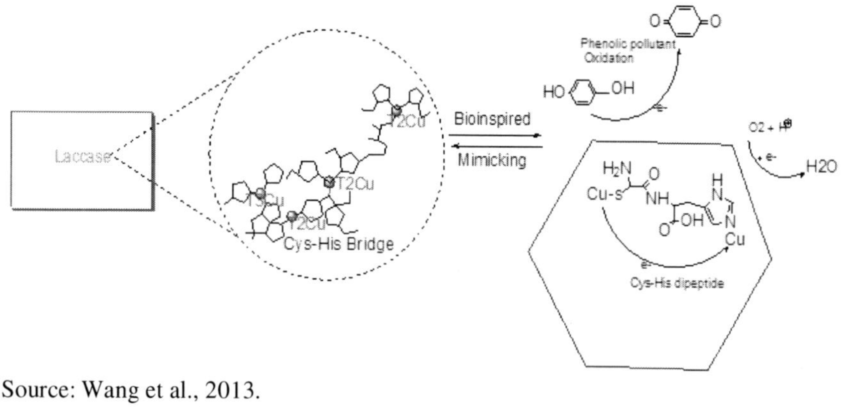

Source: Wang et al., 2013.

Figure 2. Cysteine-histidine Cu nanozyme for phenolic pollutant oxidation.

3. Functional Mimic Type Nanozymes (Type II)

There are only a few examples of type II nanozymes that have been shown to decompose organic contaminants in the environment (Table 2). The predominant activity identified was peroxidase-like activity (Jain et al., 2019; Wang et al., 2020; Safavi et al., 2012), while oxidase-like activity (Xu et al.,

2011) was also found. Fe(0) particles were explored as heterogeneous Fenton-like catalysts for the elimination of 4-chloro-3-methyl phenol. Lower pH was shown to increase catalytic activity, and the reaction was still quite efficient up to pH 6.1. LC-MS revealed a variety of intermediates and products, including chloride ion, oxalic acid, acetic acid, and formic acid among the final products after 60 minutes of reaction. The catalytic activity of the material declined over time, and its recyclability was not confirmed. In this work, they're beneficial of as a pre-catalyst. In another work, Chen et al., (2016) demonstrated that cubic boron nitride could oxidise TMB and had peroxidase-like activity. Cubic boron nitride is a material that can be reused several times. The catalytic activity was maximum at pH 5 and temperatures between 40 and 50 degrees Celsius. In the presence of H_2O_2, cubic boron nitride also catalysed the degradation of Rh B.

Table 2. Functional mimic type (type II) nanozymes in various environmental pollutants remediation

Nanozyme Type	Removal of Pollutant	Enzyme-like Function	Year	Ref.
Fe(0) Nanoparticles	4-Chloro-3-methyl phenol	Peroxidase	2011	Xu et al.
Carbon nanodots	Azo dyes (methyl red and methyl orange)	Peroxidase	2012	Safavi et al.
Cubic boron nitride	Rh B	Peroxidase	2020	Wang et al.
Graphitic carbon nitride	TMB	Dual oxidase-peroxidase	2019	Jain et al.

4. Nanocomposite Type Nanozymes (Type III)

Nanocomposite materials, such as metal/metal oxides on carbon materials, MOFs, and bimetallic alloys (core-shell), have recently been widely explored as environmental remediation agents. Composite nanozymes' primary enzymatic activity in the breakdown of organic contaminants is peroxidase-like activity. Phenolic compounds (Zuo et al., 2009) [48,158,186–189], dyes (Tial et al., 2011; Zhang et al., 2013; Zubir et al., 2014; Hu et al., 2014, Cheng et al., 2014), and organophosphorus compounds were among the refractory organic pollutants studied in these investigations.As nanocomposites with metal/metal oxide nanoparticles, various types of carbon materials have been explored. Carbon nanotubes, multiwalled carbon nanotubes (MWNTs), graphene, graphene quantum dots, graphene oxide, mesoporous carbon, and other materials are examples (Rebeiro et al., 2016). Fe_3O_4 MNP/MWNT nanocomplexes exhibiting peroxidase-like activity were synthesised by Zuo et al., (2009). The nanocomplex efficiently catalysed the oxidative degradation of phenols to insoluble polyaromatic compounds that could be easily isolated from aqueous solutions in the presence of H_2O_2. The breakdown of Orange G dye was aided by magnetically recoverable Fe_xO_y-MWNT (Variava et al., 2012). Fe_3O_4 nanoparticles produced on carbon nanotubes (7 nm in size) were found to have higher catalytic reactivity as an enzyme mimic for the breakdown of the dye Orange II than Fe_3O_4 nanoparticles.

5. 3D Structural Mimic Nanozymes (Type IV)

There are only a few studies that have used type IV nanozymes to degrade environmental contaminants. The technical difficulty in constructing these 3D nanostructures is likely to be a role in this limitation.Zhu and Diao (2011) developed porous Fe_3O_4 nanospheres and demonstrated that they were very effective as catalysts for xylenol orange degradation in aqueous solution using H_2O_2 as the oxidant. The porous Fe_3O_4 may be recycled seven times with only a minor decrease in activity after each cycle. Through the infiltration of $HAuCl_4$ into hollow Au@Ag@ICPs core-shell nanostructures and its replacement reaction with Au@Ag nanoparticles, Wang et al., (2015) created three-dimensional nano-assemblies of noble metal nanoparticle (NP)–infinite coordination polymers (ICPs). These three-dimensional nano-assemblies have oxidase-like activity. TMB was oxidised without the use of H_2O_2 to produce its blue product using surface-adsorbed O_2. The oxidase-like activity of the 3D assemblies was also used to decompose methylene blue.

Conclusion

Peroxidase-like catalytic activity has been the most extensively studied in the degradation of environmental contaminants, such as phenols, Rh B, dyes (methylene blue, acid orange), and other chemicals. Nanozymes offer a number of benefits, including low cost, ease of preparation, high stability, and recyclability. Nanozymes based on zero valent metals have been used as catalysts for environmental pollution remediation the least. Noble metals are costly in large-scale applications, whereas non-noble metals with zero valents are prone to oxidation and deterioration. Furthermore, disease detection could benefit from 3D-structured materials with excellent specificity (moulded at active regions). High selectivity and specificity are not always required in pollutant degradation. Furthermore, 3D structure fabrication currently necessitates tremendous technical effort.

References

Bhuyan, D., Arbuj, S. S., Saikia, L. Template-free synthesis of Fe3O4 nanorod bundles and their highly efficient peroxidase mimetic activity for the degradation of organic dye pollutants with H2O2. *New J. Chem.* 2015, 39, 7759–7762.

Chang, Y. H., Yao, Y. F., Luo, H., Cui, L., Zhi, L. J. Magnetic Fe3O4-GO nanocomposites as highly efficient Fenton-like catalyst for the degradation of dyes. *Int. J. Nanomanuf.* 2014, 10, 132–141.

Chen, T. M., Xiao, J.,Wang, G. W. Cubic boron nitride with an intrinsic peroxidase-like activity. *RSC Adv.* 2016, 6, 70124–70132.

Cheng, R., Li, G.-q., Cheng, C., Shi, L., Zheng, X., Ma, Z. Catalytic oxidation of 4-chlorophenol with magnetic Fe3O4 nanoparticles: Mechanisms and particle transformation (2015) *RSC Adv.* 5, 66927–66933.

Feng, Y. B., Hong, L., Liu, A. L., Chen, W. D., Li, G. W., Chen, W., Xia, X. H. High-efficiency catalytic degradation of phenol based on the peroxidase-like activity of cupric oxide nanoparticles. *Int. J. Environ. Sci. Technol.* 2015, 12, 653–660.

Harrad, S. *Persistent Organic Pollutants*; Wiley: Chichester, UK, 2010.

He, J., Liang, M. Nanozymes for environmental monitoring and treatment. In: *Nanozymology*; Yan, X., Ed.; Springer: Berlin, Germany, 2020; pp. 527–543.

Hu, P., Han, L., Dong, S. A facile one-pot method to synthesize a polypyrrole/hemin nanocomposite and its application in biosensor, dye removal, and photothermal therapy. *ACS Appl. Mater. Interfaces* 2014, 6, 500–506.

Huang, R., Fang, Z., Yan, X., Cheng, W. Heterogeneous sono-Fenton catalytic degradation of bisphenol A by Fe3O4 magnetic nanoparticles under neutral condition. *Chem. Eng. J.* 2012, 197, 242–249.

Huang, Y., Ren, J., Qu, X. Nanozymes: Classification, catalytic mechanisms, activity regulation, and applications (2019) *Chem. Rev.* 119, 4357–4412.

Hussain, C. M. *The Handbook of Environmental Remediation: Classic and Modern Techniques*; Royal Society of Chemistry: Cambridge, UK, 2020.

Jain, S., Panigrahi, A.; Sarma, T. K. Counter anion-directed growth of iron oxide nanorods in a polyol medium with efficient peroxidase-mimicking activity for degradation of dyes in contaminated water. *ACS Omega* 2019, 4, 13153–13164.

Janoš, P., Kuráˇ N, P., Pilaˇrová, V., Trögl, J., Št'astný, M., Pelant, O., Henych, J., Bakardjieva, S., Životský, O., Kormunda, M., Mazanec K., Skoumal M. Magnetically separable reactive sorbent based on the CeO2/ -Fe2O3 composite and its utilization for rapid degradation of the organophosphate pesticide parathion methyl and certain nerve agents. *Chem. Eng. J.* 2015, 262, 747–755.

Kazlauskas, R. J. Enhancing catalytic promiscuity for biocatalysis (2005) *Curr. Opin. Chem. Biol.* 9, 195–201.

Khersonsky, O., Roodveldt, C., Tawfik, D. S. Enzyme promiscuity: Evolutionary and echanistic aspects (2006) *Curr. Opin. Chem. Biol.* 10, 498–508.

Liu, Y., Zhu, G., Bao, C., Yuan, A., Shen, X. Intrinsic peroxidase-like activity of porous CuO micro-/nanostructures with clean surface. *Chin. J. Chem.* 2014, 32, 151–156.

Meng, Y., Li, W., Pan, X., Gadd, G. M. Applications of nanozymes in the environment (2020) *Environ. Sci. Nano* 2020, 7, 135–1318.

Ribeiro, R. S., Silva, A. M. T., Figueiredo, J. L., Faria, J. L., Gomes, H. T. Catalytic wet peroxide oxidation: A route towards the application of hybrid magnetic carbon nanocomposites for the degradation of organic pollutants. A review. *Appl. Catal. B Environ.* 2016, 187, 428–460.

Safavi, A., Sedaghati, F., Shahbaazi, H., Farjami, E. Facile approach to the synthesis of carbon nanodots and their peroxidase mimetic function in azo dyes degradation. *RSC Adv.* 2012, 2, 7367–7370.

Scott, C., Pandey, G., Hartley, C. J., Jackson, C. J., Cheesman, M. J., Taylor, M. C., Pandey, R., Khurana, J. L., Teese, M., Coppin, C. W., Weir K. M., Jain R. K., Lal R., Russell R. J., Oakeshott J. G. The enzymatic basis for pesticide bioremediation (2008) *Indian J. Microbiol.* 48, 65–79.

Sharma, B., Dangi, A. K., Shukla, P. Contemporary enzyme based technologies for bioremediation: A review (2018) *J. Environ. Manag.* 210, 10–22.

Tian, S. H., Tu, Y. T., Chen, D. S., Chen, X., Xiong, Y. Degradation of acid Orange II at neutral pH using Fe2(MoO4)3 as a heterogeneous Fenton-like catalyst. *Chem. Eng. J.* 2011, 169, 31–37.

Variava, M. F., Church, T. L., Harris, A. T. Magnetically recoverable FexOy–MWNT Fenton's catalysts that show enhanced activity at neutral pH. *Appl. Catal. B Environ.* 2012, 123, 200–207.

Wang, J.; Huang, R.; Qi, W.; Su, R.; Binks, B. P.; He, Z. Construction of a bioinspired laccase-mimicking nanozyme for the degradation and detection of phenolic pollutants. *Appl. Catal. B Environ.* 2019, 254, 452–462.

Wang, L., Zeng, Y., Shen, A., Zhou, X., Hu, J. Three dimensional nano-assemblies of noble metal nanoparticle-infinite coordination polymers as specific oxidase mimetics for degradation of methylene blue without adding any cosubstrate. *Chem. Commun.* 2015, 51, 2052–2055.

Wang, S., Bromberg, L., Schreuder-Gibson, H., Hatton, T. A. Organophophorous ester degradation by chromium (III) terephthalate metal–organic framework (MIL-101) chelated to N,N-dimethylaminopyridine and related aminopyridines. *ACS Appl. Mater. Interfaces* 2013, 5, 1269–1278.

Wang, S., Fang, H., Yi, X., Xu, Z., Xie, X., Tang, Q., Ou, M., Xu, X. Oxidative removal of phenol by HRP-immobilized beads and its environmental toxicology assessment (2016) *Ecotoxicol. Environ. Saf.* 130, 234–239.

Wang, Y., Liu, T., Liu, J. Synergistically boosted degradation of organic dyes by CeO2 nanoparticles with fluoride at low pH. *ACS Appl. Nano Mater.* 2020, 3, 842–849.

Wei, H. and Wang, E. K. Nanomaterials with enzyme-like characteristics (nanozymes): Next-generation artificial enzymes (2013) *Chem. Soc. Rev.* 42, 6060–6093.

Wu, J. J. X., Wang, X. Y., Wang, Q., Lou, Z. P., Li, S. R., Zhu, Y. Y., Qin, L., Wei, H. Nanomaterials with enzyme-like characteristics (nanozymes): Next-generation artificial enzymes (II) (2019) *Chem. Soc. Rev.* 48, 1004–1076.

Xu, L., Wang, J. A heterogeneous Fenton-like system with nanoparticulate zero-valent iron for removal of 4-chloro-3-methyl phenol. *J. Hazard. Mater.* 2011, 186, 256–264.

Xu, L., Wang, J. Fenton-like degradation of 2,4-dichlorophenol using Fe3O4 magnetic nanoparticles. *Appl. Catal. B Environ.* 2012, 123–124, 117–126.

Zeb, A., Xie, X., Yousaf, A. B., Imran, M., Wen, T., Wang, Z., Guo, H.-L., Jiang, Y.-F., Qazi, I. A., Xu, A.-W. Highly efficient Fenton and enzyme-mimetic activities of mixed-phase VOx nanoflakes. *ACS Appl. Mater. Interfaces* 2016, 8, 30126–30132.

Zhang, J., Zhuang, J., Gao, L., Zhang, Y., Gu, N., Feng, J., Yang, D., Zhu, J., Yan, X. Decomposing phenol by the hidden talent of ferromagnetic nanoparticles. *Chemosphere* 2008, 73, 1524–1528.

Zhang, Z., Hao, J., Yang, W., Lu, B., Ke, X., Zhang, B., Tang, J. Porous Co3O4 nanorods–reduced graphene oxide with intrinsic peroxidase-like activity and catalysis in the degradation of methylene blue. *ACS Appl. Mater. Interfaces* 2013, 5, 3809–3815.

Zhu, M., Diao, G. Synthesis of porous Fe3O4 nanospheres and its application for the catalytic degradation of xylenol orange. *J. Phys. Chem.* C 2011, 115, 18923–18934.

Zubir, N. A., Yacou, C., Motuzas, J., Zhang, X., Diniz da Costa, J. C. Structural and functional investigation of graphene oxide–Fe3O4 nanocomposites for the heterogeneous Fenton-like reaction. *Sci. Rep.* 2014, 4, 4594.

Zuo, X., Peng, C., Huang, Q., Song, S., Wang, L., Li, D., Fan, C. Design of a carbon nanotube/magnetic nanoparticle-based peroxidase-like nanocomplex and its application for highly efficient catalytic oxidation of phenols. *Nano Res.* 2009, 2, 617–623.

Chapter 6

Future Prospects towards Catalytic Defense against Microbial Pathogens Using Nanozymes

C. D. S. L. N. Tulasi, D. Manikantha, Rajesh Abhinav Bokinala and Kalyani Chepuri*

Institute of Science and Technology,
Jawaharlal Nehru Technological University Hyderabad, Hyderabad, India

Abstract

Nanozymes are a special sort of mimic enzyme that blends nanomaterial with catalytic activity. Researchers are interested in nanozymes because of their unique features, such as good stability, economy, multi-functionality, and ease of large-scale synthesis. Each year, bacterial infection-related illnesses become increasingly widespread, posing a global health threat. Antibacterial drugs that are both inventive and effective in binding to microbes are desperately needed. Antibiotics have been the most comprehensive treatment until now, but they have the potential to be overused and misused, leading to an increase in multidrug resistance.

In the current context, antibiotic usage, as well as bacterial antibiotic resistance, is a global concern. It's both desirable and necessary to develop new antimicrobial agents that kill microbes without developing resistance or biosafety problems. Nanozymes can destroy a greater number of microbes than regular enzymes, bridging the gap between biology and nanotechnology. Nanozymes are strong nanoantibiotics with broad-spectrum antibacterial properties and low biotoxicity.

This chapter highlights recent nanozyme innovations as well as modes of action against bacteria. Finally, difficulties and restrictions for

* Corresponding Author's Email: kalyanichepuri@gmail.com.

In: Emerging Environmental Applications of Nanozymes
Editors: Seema Nara and Smriti Singh
ISBN: 979-8-88697-552-9

enhancing antibacterial activity, as well as prospective approaches for using developed nanozymes with improved antibacterial function, are reviewed.

Keywords: nanozymes, multidrug resistance, catalytic activity, biotoxicity

1. Introduction

In humans, bacterial infections are the leading causes of death [1-3]. Since the discovery of antibiotics, the death rate from bacterial infections has considerably decreased, and severe diseases like tuberculosis are currently under control [4-6]. Most antibiotics can delay the growth or kill the pathogenic bacteria by interrupting their plasma membrane or cell wall formation, deoxyribonucleic acid (DNA) repair or replication and protein synthesis process [7-9]. But extensive use of antibiotics has led to the development of antimicrobial resistance (AMR), which is a serious risk to global public health and also contaminates the environment through the food chain consequently leading to the disruption of ecosystems [10, 11]. In 2019, an estimated 4.95 million people have died as a result of bacterial AMR [12, 13]. Bacteria can form biofilms, which are clusters of bacteria covered by extracellular polymeric substances (EPS). These biofilms are known to protect the bacteria through antibiotics and also the immune system of the host [14-16]. As a result, discovering novel antibacterial drugs or strategies to attack bacteria is a massive challenge.

Alternatively, broad-spectrum antibacterial drugs or techniques are the focus of substantial efforts to deal with AMR [17-19]. In this connection, some nanomaterial-based strategies like PTT (Photo-thermal therapy) [20], PDT (Photo-dynamic therapy) [21], and PCT (photo-catalytic therapy) [22] are acting as antibacterial platforms which are driving attention of scientists now a days. Furthermore, recent research has also demonstrated that the human body's inherent self-defense systems can use biocatalysts in restricting the growth of bacteria or destroying the biofilms [23-25]. Enzymes such as natural oxidase (OXD) and peroxidase (POD) in living bodies, can catalyze a variety of components in producing ROS (reactive oxygen species) that could be used in fighting bacterial invasion. As a result of their high selectivity and great catalytic activity for certain substrates, enzymes are preferable alternatives to antibiotics. Natural enzymes, on the other hand, have various drawbacks that drastically reduce their ability as antibacterial agents, such as high manufacturing costs, low catalytic stability, and complex production conditions [26]. Though being more membrane permeable and biocompatible than conventional antibiotics, nanozymes can minimize the emergence of bacterial resistance (Figure 1) [27-30].

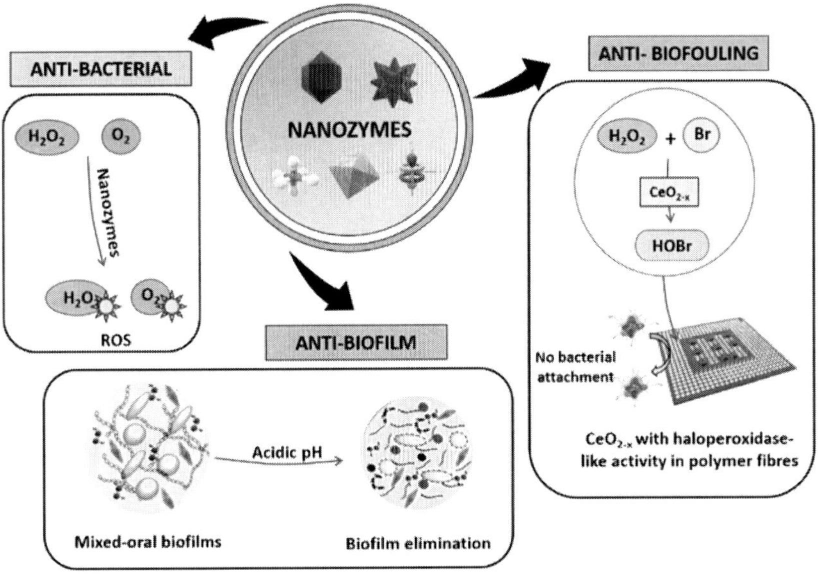

Figure 1. Flow diagram showing antibacterial nanozymes and their applications.

Nanozymes have emerged as a research hotspot for fighting bacteria due to their high stability, simplicity of manufacturing, low cost and multi-functionality [31-33]. Additionally, nanozymes have characteristics beyond those of natural enzymes due to their distinct physicochemical properties to modify catalytic activity, composition, size, and shape [38, 39]. These unique physicochemical properties raise the prospect of developing multifunctional antibacterial agents [34-37]. Metal based compounds, transition metal peroxides/oxides/dichalcogenides, carbon-based nanomaterials, SAzymes or single atom nanozymes, as well as MOFs or metalorganic frameworks based compounds are some of the examples of nanozymes that have been developed till date to battle AMR [40-42].

2. Nano-Enzyme Mediated Antibacterial Mechanisms

The development of biomedical technology requires a thorough understanding of the mechanisms underlying nanozyme-mediated antibacterial action. Despite significant efforts, universal mechanism for nanozyme antibacterial role is yet to be properly explored and fully understood due to the complexity of nanozyme categories, physicochemical features, and interference concerns. Three primary mechanisms for the antibacterial actions of nanozymes are summarized based on recent findings: production of HOBr/Cl, ROS modulation, and clearance of extracellular DNA.

2.1. ROS Regulation

Reactive oxidative species (ROS) are intermediate chemical species formed during the incomplete oxygen reduction process. They are also formed during the oxidative metabolism that occurs in mitochondria. These are highly reactive and quite unstable due to the presence of one or more unpaired electrons. Hydrogen peroxides (H_2O_2), hydroxyl radicals (OH), superoxide anion (O_2^-), and single oxygen ($^1O_2$) are examples of ROS [43].

On one hand, excessive production of ROS in disease conditions like cancer, inflammation, diabetes, aging, etc., [44] can cause damage to cellular components such as DNA, RNA, proteins, etc. and on the other hand these ROS are used by our immune system to fight infections. This has led to a general interest in the research of ROS concentration regulation, notably in ROS-regulating therapies such ROS production triggered toxic therapy and the ROS elimination triggering antioxidant therapy.

Generally, the exogenous ROS is responsible for inhibiting the bacterial growth by inducing irrepairable damage to the DNA, cell wall, plasma membrane, proteins, nucleic acids, and polysaccharides [45]. Most significantly, ROS can inhibit bacterial biofilm formation or even degrade existing biofilms. Oxidases and Peroxidases, which are influenced by natural enzymes, are main design concepts for nanozymes in the production of ROS. In general, Hydrogen peroxide (H_2O_2) has inherent antibacterial properties at higher concentrations (0.5 to 3%, or 166 mM to 1000 mM), whilst normal tissues could get damaged at these high concentrations. Interestingly, Peroxidase like nanozymes have been reported in literature, which catalyze the conversion of modest concentrations of H_2O_2 (<1 mM) to extremely toxic OH, that can effectively kill bacteria [46]. In assessing the nanozyme-substrate affinity and catalytic activity, the enzyme kinetic parameters, like Vmax (maximum initial velocity) and Km (Michaelis Menten constant) are determined by adjusting the substrate's (TMB tetramethylbenzidine or H_2O_2) initial concentration [47]. These values were used in determining whether the reaction fits the equation of Michaelis-Menten (Figure 2).

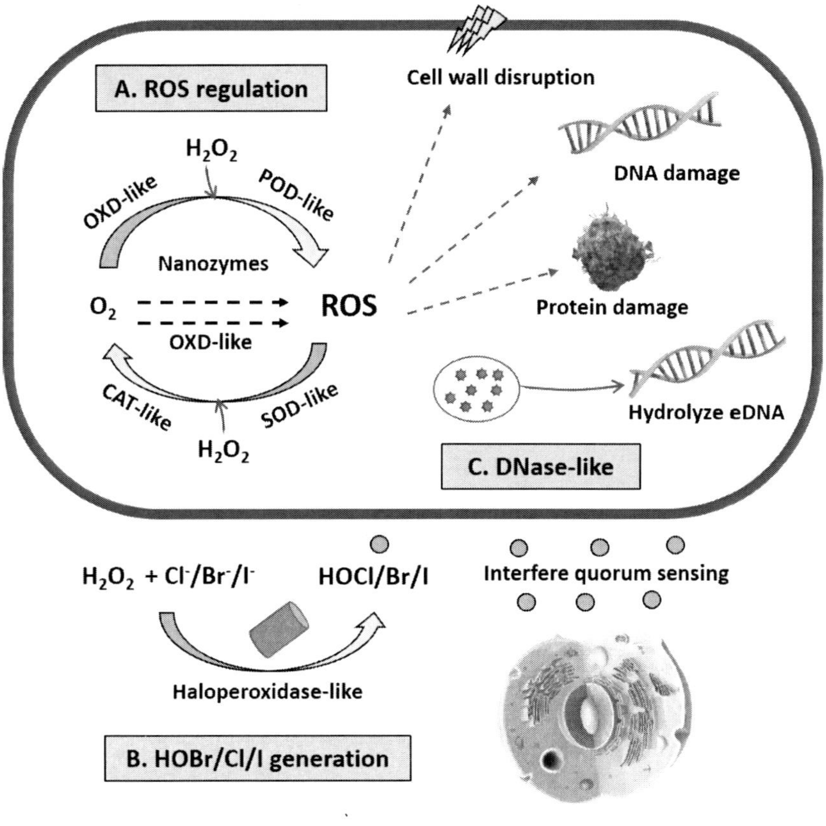

Figure 2. Nanozymes' antibacterial activities are mediated by three mechanisms: a) Regulation of (ROS) for controlling bacterial inflammation or organelle damage (such as cell walls, protein, and DNA). b) Interfering quorum sensing with HOBr/Cl generation, and c) DNase-like nanozymes remove extracellular DNA.

2.2. HOBr/Cl Production

Biofouling, a primary contributor to pollution in the environment, is mostly generated by quorum sensing (QS) bacteria. This determines the cell-to-cell communication method which allows the bacteria in coordinating the cluster behavior throughout the biofilm formation, bioluminescence, and virulence processes [48]. Marine algae possess strong anti-biofouling ability as they are known to release haloperoxidases. The three types of haloperoxidases are chloroperoxidase, bromoperoxidase, and iodoperoxidase. They play a role in catalyzing halide (Br, Cl, and I) reaction with H_2O_2 to produce hypohalous acid (HOBr, HOI, and HOCl). Hypohalous acid, which plays a key role in biological anti-fouling, can cause irreversible damage to microorganisms and biofilms. Interestingly, some cerium and vanadium-based nanozymes have demonstrated strong haloperoxidase-activity in Br oxidation into HOBr and $^1O_2$ in presence of H_2O_2. It prevents the formation of biofoul through quenching auto-inducers like N-acyl homoserine lactones and furanosyl compounds [49]. Furthermore, haloperoxidase-like nanozymes are usually not biocompatible, but catalytic stability is maintained extensively into a hostile environment.

2.3. Extracellular DNA (eDNA) Clearance

The extra cellular DNA proteins, and lipopolysaccharide are involved in the formation of Extra cellular matrixes (ECM) and biofilms to shield bacteria from the antibiotics. Extra cellular DNA (eDNA), the most important components of the ECM is responsible for bacterial attachment on substrate surface and bacteria to bacteria interconnector, ensuring biofilm's integrity [50]. Natural deoxyribonuclease (DNase) has been presented as a way to remove ECM and prevent biofilm formation. Natural DNase, on the other hand, has several limitations, including expensive, irreversibility, and short-term catalytic activity, which severely limits the utility as anti-bacterial agent against the biofilms.

In mitigating the problem, many nanozymes possessing DNase like activity have been already reported to destroy biofilms by speeding up DNA lysis in presence of water (hydrolysis) [51]. When compared to natural DNase, nanozymes offer unparalleled prospects as antibiofilm agents due to their variable catalytic activity, strong recyclability and cost-effectiveness. Furthermore, disturbing the compactness of ECM in the presence of DNase nanozymes dramatically boosts the efficacy of standard antibiotics in eradicating the confined bacteria, which could be a potential technique for combating drug-resistant bacteria in the long run.

Table 1. Summary of nanozymes for antibacterial and anti-biofilms applications

S.No.	Nanozymes	Types of Bacteria	Mimicked Enzymes	Antibacterial Mechanism	Ref.
1.	Graphene quantum dots	*S. aureus* *E. coli*	Peroxidase	Release of ·OH	[60]
2.	Gold NPs	*S. aureus* *E. coli*	Peroxidase and oxidase	Releaseof $\cdot O_2^-$ and ·OH	[61]
3.	Au DAMP	*E. coli*, MDR *K. pneumoniae*, *A. baumannii*, *P. aeruginosa*, MRSA	Peroxidase and oxidase	Release of $\cdot O_2^-$ and ·OH	[62]
4.	Pd@Pt-T_{790}	*MRSA-induced myositis*	Catalase	Release of $^1O_2$, sonodynamic therapy	[63]
5.	Copper/Carbon	*S. aureus, E. coli*	Peroxidase, catalase and superoxide	Release of ·OH, release Cu^{2+}	[64]
6.	Cu_2WS_4	*S. aureus, E. coli*	Peroxidase and oxidase	Release of H_2O_2 and · OH	[65]
7.	IONzymes	*S. mutans*	Peroxidase	Release of H_2O_2 and ·OH	[66]
8.	Co-V MMO nanowires	*E. coli*	Peroxidase and oxidase	Release of $\cdot O_2^-$	[67]
9.	Cit-MoS (III)	*S. aureus, E. coli*	Peroxidase	Release of ·OH, Conversion of surface charge	[68]
10.	CeO_{2-x} nanorod	Biofilms	Haloperoxidase	HOBr release for the signal molecule bromination	[69]
11.	Tb_4O_7 NPs	*S. aureus*, *E. coli*	Oxidase	Consume antioxidant biomolecules, release of ROS	[70]
12.	V_2O_5 nanowires	Biofilms	Haloperoxidase	HOBr release for the signal molecule bromination	[71]
13.	Co_4S_3/Co_3O_4	*E. coli, S. sciuri*	Peroxidase and oxidase	Release of $\cdot O_2^-$	[72]
14.	DMAE	Biofilms	DNase	Multivalent Ce^{IV} centre for hydrolysing eDNA, Release of ·OH	[73]
15.	N-SCSs	*S. aureus, E. coli*	Peroxidase, catalase, oxidase and superoxide dismutase	Combinational of the release of ·OH and photothermal therapy	[74]

3. Factors and Optimization Measures That Influence Catalytic Activity of Nanozymes

The majority of the naturally existing enzymes use complex cofactors to perform redox and electron transfer functions Although nanozymes have distinct benefits over natural enzymes, their catalytic activity is strongly influenced by a variety of parameters like pH, size, shape, temperature, dopant, light, surface modification, etc. [52].

Nanozymes, have displayed, size-dependent catalytic activity. Smaller nanozymes can attach better to substrates and can also penetrate through bacterial cell membranes when compared to their larger counterparts. Hu et al. created gold nanoparticles of diameter 1.96 nm, ultrasmall in size [53]. These Au NPs showed exceptional POD-like action as well as extremely toxic OH production for killing bacteria under *in vitro* and *in vivo* conditions. Similarly, due to their various interactions with bacteria, nanozymes can have a significant influence on the catalytic and antimicrobial capacities. Puvvada et al. for example, discovered that when compared with sphere-shaped NPs, 8 headed (octahedron) shaped magnetite nanoparticles have stronger mimetic activity of POD (Peroxidase) [54].

Besides size as an influencing factor, both temperature and pH of the reaction media has a significant impact on nanozyme's catalytic activity. Natural enzymes' catalytic activity is significantly reduced at low pH and high temperatures; however, nanozymes are extremely resistant to these conditions. For example, when a wide pH range of 2.0-7.5 is considered, Nanosheets of MoS_2 have demonstrated strong POD-like activity. In addition, the temperature influence on POD-like nanozyme activity is studied. Lu et al. discovered, Fe_3O_4 nanoparticles with a diameter of 300 nm and at temperatures between 30°C and 55°C, maintained high POD-like catalytic activity, but (HRP) horseradish peroxidase has lost its activity when temperature was elevated beyond 30°C. The enzyme-like forms of nanozymes are also affected by pH levels. Chen et al. proposed that the CAT-(Catalase) like activity possessed by iron oxide nanoparticles at neutral pH and POD-like activity at the low pH [55].

Furthermore, nanozymes catalytic activity is linked to the Bio-nano interaction. Various topographical topologies of composites of the nanozymes could improve bacterial adherence [56]. Nanozymes' activity can also be boosted by oxygen vacancies. For fighting multidrug-resistant bacterial infections, Zhang et al. produced MoO_3xNDs (oxygen-vacancy molybdenum trioxide) which possessed improved POD-like activity [57]. The Km and Vmax of MoO_3xNDs using H_2O_2 as a substrate was 0.26 mM, which was estimated to be 14 times lesser than HRP and 1.7 times that of HRP respectively. These findings suggested that the POD-like activity of MoO_3 x NDs is high due to an abundance of oxygen-vacancy.

Several studies found that using light's great temporal and spatial precision, light expedited energy and electron transit between substrates, nanozymes are improving their catalytic efficacy. Most light-activated nanozymes have so far been focused on the formation of reactive oxygen species (ROS), but there have also been attempts to mimic other enzymes. Xi et al. generated albumin-stabilized CuS NPs by adding a photoacid [58].

Finally, recent research has shown that adenosine triphosphate (ATP), metal ions, vitamins, melamine, phosphates, and various other tiny compounds in nanozymes are considered as cofactors in catalytic reactions by influencing energy and electron transfer. Chishti et al. used co-precipitation methods to prepare cerium oxide nanocrystals (CeO_2) with a Ce^{3+} fraction of 23.04% [59]. It was observed that increasing the affinity between H_2O_2 and CeO_2 with ATP improves the POD like activity and CeO_2 nanocrystal's antibacterial characteristics at the neutral pH (Table 1).

4. Antibacterial Nanozymes: Biosafety Profiles

Biosafety, which relates to the impact of nanozymes on human health and the environment, is essential in the conversion process that turns nanomaterials into their therapeutic uses. Besides having large and optimistic benefits in antibacterial applications, it is also crucial to take into consideration, the biosafety issues. Antibacterial nanozymes, in particular, are intended to be ingested and interact with the cells and the living tissues. Although, most research concentrates on antibacterial properties of the nanozymes while ignoring their biosafety. As a result, more systematic evaluations of nanozyme's biosafety, such as absorption, biodistribution, clearing mechanism, metabolism, long term and acute toxicity, pharmacokinetics, and therapeutic effect duration, are needed in authenticating the activities that affect humankind at a low rate and do not cause adverse effects in the ecological system [75].

Because small components are more accepted in biological applications, nanomaterial selection is critical. Even though metal based nanozymes have significant bactericidal properties, the metal ion release possesses a health risk.

Fe^{3+}, Zn^{2+}, and Cu^{2+} are among the metal ions that can affect inactivation of protein substantially, which can lead to metal toxicity in many forms from the cellular to the organ level. According to Zhang et al. copper oxide nanoparticles may lead to cell mortality in HUVEC (human umbilical vein endothelial cells) as high amount of copper ions is accumulated inside the lysosome [76]. CuO NPs caused mitochondrial dysfunction and autophagy in another investigation by creating excessive superoxide ions, which resulted in cell lysis. More crucially, some investigations have found the metal ion induced amyloid beta (A) peptide aggregation, cholinergic disproportion. It led to pro inflammatory response, neurotoxicity and also plays a role in Alzheimer's disease. Although, AuNPs are antibacterial nanozymes, the potential toxicity assessment in humans is still inconclusive. Au NPs were found to be harmless to humans in certain studies, but other experimental examinations proved their harmful effects both *in vitro* and *in vivo* [77]. Because of their biosafety, carbon-based nanozymes have gained a lot of consideration as metal-free nano-catalysts for biological processes [78].

The shape, size, surface charge, composition, dosage, and appropriate functional groups of nanozymes have been shown to have a significant impact on their specificity, toxicity, as well as their catalytic activity, in several studies. Enhancing the size of nanoparticle can increase the efficacy of bacterial lysis but also increasing the toxicity to healthy cells and organs [79]. It's yet unclear how specific parameters influence nanozyme toxicity. So, more research in harmful processes for nanozymes is needed. According to some bio distribution studies, nanozymes have been found to collect mostly in the liver, lung, and spleen in the absence of targeting molecules. Immobilizing nanozymes in polymeric nanocarrier matrix or functionalizing with surface coatings like dextran, or poly-ethylene glycol, or chitosan is another way to reduce their toxicity. Nanozymes must also be endowed with the ability to target via a surface modification to improve therapeutic efficacy and reduce the cytotoxicity. Zhao et al. made AuPt bimetallic nanoparticles that were selectively harmful to the bacterial cells but not for the mammalian cells [80].

In order to comply with the biosafety regulations of nanozymes, complex modifications in biological systems are always made, and these changes must be verified over an extended period of time. To reduce the toxicity of nanozymes, the following measures may be considered. For starters, nanozymes with an ultra-small sizes and activity at high range are recommended since these may be excreted from the body quickly via the kidneys [81]. Second, biodegradable biocompatible compounds can be added to the surface of nanozymes to reduce nonspecific tissue uptake [82]. Finally, more studies are needed to develop new, safe, and effective antibacterial nanozymes.

Furthermore, establishing a comprehensive safety evaluation methodology for nanozymes is a necessary precondition for their clinical use [83]. Many studies have been conducted to date using cell and animal models to assess the biosafety of antibacterial nanozymes [84]. To investigate cellular absorption, distribution, and cytotoxicity mechanisms of nanozymes *in vitro*, a wide range of cell lines like normal and cancer cells were employed. MTS (3-(4,5-dimethylthiazol2-yl)-5-(3-carboxymethoxyphenyl)-2-(4-sulfophenyl) -2H-tetrazoliumsodium salts), CCK-8 (Cell Counting Kit-8), LDH (lactate dehydrogenase), MTT (methyl thiazolyl tetrazolium) assays are all common toxic-testing procedures. The viability of nanozyme treated cells is mostly influenced by the dosage and incubation period. Added to it, the nanozyme toxicology studies were conducted on mice, studying behavior, biochemistry, body weight, serum and histology [85]. Nanozyme toxicity testing in higher mammals, on the other hand, is still uncertain. In addition, thorough toxicity studies incorporating particle form, size, and surface chemistry are necessary to ensure the nanozyme's appropriateness for biological applications *in vivo*. Further research into bio-nano interactions, neurotoxicity, genotoxicity and immunotoxicity of nanozymes and from molecular to organism level is still in its early stages and should be thoroughly investigated. Since the nanomaterials were designed to interact with the cells, they must not have any negative consequences in the human body.

Future Perspectives and Conclusion

In this chapter, a detailed note on recent advances in anti-bacterial nanozymes and applications in fighting bacteria, removing biofilms, and treatment for environmental biofouling was elaborated. Nanozymes are approved as most promising antibacterial agents as they possess distinct advantages over natural enzymes, including high stability, cost-effectiveness, customizable shape and size, ease in preparing, and the high catalytic activity. Nanozymes having potent catalytic activity are able to produce toxic free radicals at higher quantities, which can damage components of cell/biofilms or impair bacterial DNA synthesis, resulting in improved bacterial lysis. Although existing nanozymes have strong antibacterial properties, the majority of their bactericidal applications are still in initial levels of expansion. Due to the current lack of trials, there is not much progress in nanozymes for clinical translation.

Initially, nanozyme's catalytic activity is still significantly lower than that of natural enzymes, making them unsuitable for *in vivo* antibacterial applications. Recent research has shown that nanozyme catalytic activity is influenced by both intrinsic physicochemical qualities (such as size, form and surface modification) and the extrinsic variables (pH, temperature, substrate concentration).

Moreover, there is limited knowledge on how the complex biological milieu influences nanozyme catalytic activity and its long-term consequences. As a result, it is crucial in to develop efficient nanozymes with high catalytic activity in the biological systems. Scientists should continue to work on developing appropriate supporting materials for antibacterial nanozymes with high biocompatibility and high catalyst loading capability. Nanozymes that are similar to natural metallo enzymes and have atomically scattered active sites and are responsible for significant advances in efficient catalysis.

Secondly, some nanozymes possess limited selectivity to the specific substrate due to their multienzyme-like activity, causing numerous side effects in the biological contexts. As a result, nanozyme-based precision therapy would be a popular topic in future and massive efforts are necessary in improving nanozyme selectivity for the targeted therapy.

Furthermore, most nanozymes may limit bacterial growth *in vitro*. However, *in vivo* models and treatment techniques are in the early stages. Given biological system's complexity, expanding nanozyme in vivo research is good way to assess their biocompatibility, therapeutic efficacy, and biodegradability. Nanozymes's biodistribution, *in vivo* translocation, metabolic pathways and degradation is also important for clinical translation. As a result, more work is needed to improve nanozyme biocompatibility and analyze their antibacterial efficacy *in vivo* and development of nanozyme based health devices in the future.

Finally, despite the fact that some researchers have put in credible hypotheses for the catalytic mechanisms of nanozymes, there are still many challenges to be overcome in this area. Particularly, advancements in the field of computer science such as in the molecular dynamics simulations, made it possible to learn more about the catalytic mechanisms and structure-activity relationship. As a result, at the molecular level, collaborative computational modelling, theoretical calculation, and artificial intelligence tools can clarify the catalytic mechanisms of nanozymes.

In conclusion, antibacterial nanozymes have advanced rapidly. However, none of them have yet met the standards of clinical applications. To achieve more influencing development in the future, additional efforts on outstanding issues are essential. We anticipate that this new perspective will not only pick interest of researchers onto antibacterial nanozymes, but also provide useful insights into previously undiscovered nanozyme catalytic mechanisms. Simultaneously, we hope more research could be done in developing novel kinds of antibacterial nanozymes which are safer for the future clinical use.

References

[1] Cattoir, V., and Felden, B. (2019). Future antibacterial strategies: from basic concepts to clinical challenges. *The Journal of Infectious Diseases*, *220*(3), 350-360.

[2] Weinstein, R. A., Gaynes, R., Edwards, J. R., and National Nosocomial Infections Surveillance System. (2005). Overview of nosocomial infections caused by gram-negative bacilli. *Clinical Infectious Diseases*, *41*(6), 848-854.

[3] Al-Anazi, K. A., and Al-Jasser, A. M. (2014). Infections caused by Acinetobacter baumannii in recipients of hematopoietic stem cell transplantation. *Frontiers in Oncology*, *4*, 186.

[4] Wu, Y., Song, Z., Wang, H., and Han, H. (2019). Endogenous stimulus-powered antibiotic release from nanoreactors for a combination therapy of bacterial infections. *Nature Communications*, *10*(1), 1-10.

[5] Andersson, D. I. (2015). Improving predictions of the risk of resistance development against new and old antibiotics. *Clinical Microbiology and Infection*, *21*(10), 894-898.

[6] Ding, X., Wang, A., Tong, W., and Xu, F. J. (2019). Biodegradable antibacterial polymeric nanosystems: a new hope to cope with multidrug-resistant bacteria. *Small*, *15*(20), 1900999.

[7] Kohanski, M. A., Dwyer, D. J., Hayete, B., Lawrence, C. A., and Collins, J. J. (2007). A common mechanism of cellular death induced by bactericidal antibiotics. *Cell*, *130*(5), 797-810.

[8] Kohanski, M. A., Dwyer, D. J., and Collins, J. J. (2010). How antibiotics kill bacteria: from targets to networks. *Nature Reviews Microbiology*, *8*(6), 423-435.

[9] Ter Kuile, B. H., and Hoeksema, M. (2018). Antibiotic killing through incomplete DNA repair. *Trends in Microbiology*, *26*(1), 2-4.

[10] Arias, C. A., and Murray, B. E. (2009). Antibiotic-resistant bugs in the 21st century—a clinical super-challenge. *New England Journal of Medicine*, *360*(5), 439-443.

[11] Li, Y. (2014). China's misuse of antibiotics should be curbed. *BMJ*, *348*.

[12] Ragheb, M. N., Thomason, M. K., Hsu, C., Nugent, P., Gage, J., Samadpour, A. N., and Merrikh, H. (2019). Inhibiting the evolution of antibiotic resistance. *Molecular Cell*, *73*(1), 157-165.

[13] MacLean, R. C., Hall, A. R., Perron, G. G., and Buckling, A. (2010). The population genetics of antibiotic resistance: integrating molecular mechanisms and treatment contexts. *Nature Reviews Genetics*, *11*(6), 405-414.

[14] Medzhitov, R. (2007). Recognition of microorganisms and activation of the immune response. *Nature*, *449*(7164), 819-826.

[15] Wang, L. S., Gupta, A., and Rotello, V. M. (2016). Nanomaterials for the treatment of bacterial biofilms. *ACS Infectious Diseases*, *2*(1), 3-4.

[16] Mendoza, R. A., Hsieh, J., and Galiano, R. D. (2019). The impact of biofilm formation on wound healing. *Wound Healing-Current Perspectives, 10.*

[17] Mei, L., Zhu, S., Yin, W., Chen, C., Nie, G., Gu, Z., and Zhao, Y. (2020). Two-dimensional nanomaterials beyond graphene for antibacterial applications: current progress and future perspectives. *Theranostics*, *10*(2), 757.

[18] Sun, W., and Wu, F. G. (2018). Two-Dimensional Materials for Antimicrobial Applications: Graphene Materials and Beyond. *Chemistry– An Asian Journal*, *13*(22), 3378-3410.

[19] H. Miao, Z. Teng, C. Wang, H. Chong, G. Wang, Recent progress in two[1]dimensional antimicrobial nanomaterials, *Chemistry– An Asian Journal* 25 (2019) 929–944.

[20] Ray, P. C., Khan, S. A., Singh, A. K., Senapati, D., and Fan, Z. (2012). Nanomaterials for targeted detection and photothermal killing of bacteria. *Chemical Society Reviews*, *41*(8), 3193-3209.

[21] Hamblin, M. R. (2016). Antimicrobial photodynamic inactivation: a bright new technique to kill resistant microbes. *Current Opinion in Microbiology*, *33*, 67-73.

[22] Xi, J., Wei, G., Wu, Q., Xu, Z., Liu, Y., Han, J., and Gao, L. (2019). Light-enhanced sponge-like carbon nanozyme used for synergetic antibacterial therapy. *Biomaterials Science*, *7*(10), 4131-4141.

[23] Zhao, S., Li, S., and Wei, H. (2020). Beyond: Novel Applications of Nanozymes. In *Nanozymology* (pp. 545-555). Springer, Singapore.

[24] Wang, Y., Yang, Y., Shi, Y., Song, H., and Yu, C. (2020). Antibiotic-free antibacterial strategies enabled by nanomaterials: progress and perspectives. *Advanced Materials*, *32*(18), 1904106.

[25] Liu, X., Gao, Y., Chandrawati, R., and Hosta-Rigau, L. (2019). Therapeutic applications of multifunctional nanozymes. *Nanoscale*, *11*(44), 21046-21060.

[26] Zhang, R., Fan, K., and Yan, X. (2020). Nanozymes: created by learning from nature. *Science China Life Sciences*, *63*(8), 1183-1200.

[27] Wu, J., Li, S., and Wei, H. (2018). Integrated nanozymes: facile preparation and biomedical applications. *Chemical Communications*, *54*(50), 6520-6530.

[28] Zhang, Y., Jin, Y., Cui, H., Yan, X., and Fan, K. (2020). Nanozyme-based catalytic theranostics. *RSC Advances*, *10*(1), 10-20.

[29] Cormode, D. P., Gao, L., and Koo, H. (2018). Emerging biomedical applications of enzyme-like catalytic nanomaterials. *Trends in Biotechnology*, *36*(1), 15-29.

[30] Wei, H., and Wang, E. (2013). Nanomaterials with enzyme-like characteristics (nanozymes): next-generation artificial enzymes. *Chemical Society Reviews*, *42*(14), 6060-6093.

[31] Meng, X., Fan, K., and Yan, X. (2019). Nanozymes: an emerging field bridging nanotechnology and enzymology. *Science China Life Sciences*, *62*(11), 1543-1546.

[32] Fan, K., Xi, J., Fan, L., Wang, P., Zhu, C., Tang, Y., and Gao, L. (2018). *In vivo* guiding nitrogen-doped carbon nanozyme for tumor catalytic therapy. *Nature Communications*, *9*(1), 1-11.

[33] Wang, X., Hu, Y., and Wei, H. (2016). Nanozymes in bionanotechnology: from sensing to therapeutics and beyond. *Inorganic Chemistry Frontiers*, *3*(1), 41-60.

[34] Gao, L., Giglio, K. M., Nelson, J. L., Sondermann, H., and Travis, A. J. (2014). Ferromagnetic nanoparticles with peroxidase-like activity enhance the cleavage of biological macromolecules for biofilm elimination. *Nanoscale*, *6*(5), 2588-2593.

[35] Liang, M., Wang, Y., Ma, K., Yu, S., Chen, Y., Deng, Z., and Wang, F. (2020). Engineering inorganic nanoflares with elaborate enzymatic specificity and efficiency for versatile biofilm eradication. *Small*, *16*(41), 2002348.

[36] Yang, D., Chen, Z., Gao, Z., Tammina, S. K., and Yang, Y. (2020). Nanozymes used for antimicrobials and their applications. *Colloids and Surfaces B: Biointerfaces*, *195*, 111252.

[37] Meng, Y., Li, W., Pan, X., and Gadd, G. M. (2020). Applications of nanozymes in the environment. *Environmental Science: Nano*, *7*(5), 1305-1318.

[38] Dong, H., Fan, Y., Zhang, W., Gu, N., and Zhang, Y. (2019). Catalytic mechanisms of nanozymes and their applications in biomedicine. *Bioconjugate Chemistry*, *30*(5), 1273-1296.

[39] Zhang, R., Zhou, Y., Yan, X., and Fan, K. (2019). Advances in chiral nanozymes: a review. *Microchimica Acta*, *186*(12), 1-12.

[40] Xiong, X., Huang, Y., Lin, C., Liu, X. Y., and Lin, Y. (2019). Recent advances in nanoparticulate biomimetic catalysts for combating bacteria and biofilms. *Nanoscale*, *11*(46), 22206-22215.

[41] Huang, Y., Ren, J., and Qu, X. (2019). Nanozymes: classification, catalytic mechanisms, activity regulation, and applications. *Chemical Reviews*, *119*(6), 4357-4412.

[42] Song, W., Zhao, B., Wang, C., Ozaki, Y., and Lu, X. (2019). Functional nanomaterials with unique enzyme-like characteristics for sensing applications. *Journal of Materials Chemistry B*, *7*(6), 850-875.

[43] Yang, B., Chen, Y., and Shi, J. (2019). Reactive oxygen species (ROS)-based nanomedicine. *Chemical Reviews*, *119*(8), 4881-4985.

[44] Zhang, C., Wang, X., Du, J., Gu, Z., and Zhao, Y. (2021). Reactive Oxygen Species-Regulating Strategies Based on Nanomaterials for Disease Treatment. *Advanced Science*, *8*(3), 2002797.

[45] Jiang, D., Ni, D., Rosenkrans, Z. T., Huang, P., Yan, X., and Cai, W. (2019). Nanozyme: new horizons for responsive biomedical applications. *Chemical Society Reviews*, *48*(14), 3683-3704.

[46] Attar, F., Shahpar, M. G., Rasti, B., Sharifi, M., Saboury, A. A., Rezayat, S. M., and Falahati, M. (2019). Nanozymes with intrinsic peroxidase-like activities. *Journal of Molecular Liquids*, *278*, 130-144.

[47] Lou, Z., Zhao, S., Wang, Q., and Wei, H. (2019). N-doped carbon as peroxidase-like nanozymes for total antioxidant capacity assay. *Analytical Chemistry*, *91*(23), 15267-15274.

[48] Yang, N., Zhu, M., Xu, G., Liu, N., and Yu, C. (2020). A near-infrared light-responsive multifunctional nanocomposite hydrogel for efficient and synergistic antibacterial wound therapy and healing promotion. *Journal of Materials Chemistry B*, *8*(17), 3908-3917.

[49] Hu, M., Korschelt, K., Viel, M., Wiesmann, N., Kappl, M., Brieger, J., and Tremel, W. (2018). Nanozymes in nanofibrous mats with haloperoxidase-like activity to combat biofouling. *ACS Applied Materials and Interfaces*, *10*(51), 44722-44730.

[50] Chen, Z., Wang, Z., Ren, J., and Qu, X. (2018). Enzyme mimicry for combating bacteria and biofilms. *Accounts of Chemical Research*, *51*(3), 789-799.

[51] Chen, Z., Ji, H., Liu, C., Bing, W., Wang, Z., and Qu, X. (2016). A multinuclear metal complex based DNase-mimetic artificial enzyme: matrix cleavage for combating bacterial biofilms. *AngewandteChemie*, *128*(36), 10890-10894.

[52] Kuah, E., Toh, S., Yee, J., Ma, Q., and Gao, Z. (2016). Enzyme mimics: advances and applications. *Chemistry–A European Journal*, *22*(25), 8404-8430.

[53] Hu, W. C., Younis, M. R., Zhou, Y., Wang, C., and Xia, X. H. (2020). In situ fabrication of ultrasmall gold nanoparticles/2D MOFs hybrid as nanozyme for antibacterial therapy. *Small*, *16*(23), 2000553.

[54] Puvvada, N., Panigrahi, P. K., Mandal, D., and Pathak, A. (2012). Shape dependent peroxidase mimetic activity towards oxidation of pyrogallol by H 2 O 2. *RSC Advances*, *2*(8), 3270-3273.

[55] Chen, Z., Yin, J. J., Zhou, Y. T., Zhang, Y., Song, L., Song, M., and Gu, N. (2012). Dual enzyme-like activities of iron oxide nanoparticles and their implication for diminishing cytotoxicity. *ACS Nano*, *6*(5), 4001-4012.

[56] Zhao, T., Chen, L., Wang, P., Li, B., Lin, R., Abdulkareem Al-Khalaf, A., and Zhao, D. (2019). Surface-kinetics mediated mesoporous multipods for enhanced bacterial adhesion and inhibition. *Nature Communications*, *10*(1), 1-10.

[57] Zhang, Y., Li, D., Tan, J., Chang, Z., Liu, X., Ma, W., and Xu, Y. (2021). Near-infrared regulated nanozymatic/photothermal/photodynamic triple-therapy for combating multidrug-resistant bacterial infections via oxygen-vacancy molybdenum trioxide nanodots. *Small*, *17*(1), 2005739.

[58] Xi, J., Zhang, J., Qian, X., An, L., and Fan, L. (2020). Using a visible light-triggered pH switch to activate nanozymes for antibacterial treatment. *RSC Advances*, *10*(2), 909-913.

[59] Chishti, B., Fouad, H., Seo, H. K., Alothman, O. Y., Ansari, Z. A., and Ansari, S. G. (2020). ATP fosters the tuning of nanostructured CeO 2 peroxidase-like activity for promising antibacterial performance. *New Journal of Chemistry*, *44*(26), 11291-11303.

[60] Sun, H., Gao, N., Dong, K., Ren, J., and Qu, X. (2014). Graphene quantum dots-band-aids used for wound disinfection. *ACS Nano*, *8*(6), 6202-6210.

[61] Tao, Y., Ju, E., Ren, J., and Qu, X. (2015). Bifunctionalized mesoporous silica-supported gold nanoparticles: intrinsic oxidase and peroxidase catalytic activities for antibacterial applications. *Advanced Materials*, *27*(6), 1097-1104.

[62] Zheng, Y., Liu, W., Qin, Z., Chen, Y., Jiang, H., and Wang, X. (2018). Mercaptopyrimidine-conjugated gold nanoclusters as nanoantibiotics for combating multidrug-resistant superbugs. *Bioconjugate Chemistry*, *29*(9), 3094-3103.

[63] Sun, D., Pang, X., Cheng, Y., Ming, J., Xiang, S., Zhang, C., and Zheng, N. (2020). Ultrasound-switchable nanozyme augments sonodynamic therapy against multidrug-resistant bacterial infection. *ACS Nano*, *14*(2), 2063-2076.

[64] Xi, J., Wei, G., An, L., Xu, Z., Xu, Z., Fan, L., and Gao, L. (2019). Copper/carbon hybrid nanozyme: tuning catalytic activity by the copper state for antibacterial therapy. *Nano Letters*, *19*(11), 7645-7654.

[65] J. Shan, X. Li, K. Yang, W. Xiu, Q. Wen, Y. Zhang, L. Yuwen, L. Weng, Z. Teng, L. Wang, Efficient Bacteria Killing by Cu_2WS_4 Nanocrystals with Enzyme-like Properties and Bacteria-Binding Ability, *ACS Nano* 13 (2019) 13797–13808.

[66] Wang, Y., Shen, X., Ma, S., Guo, Q., Zhang, W., Cheng, L., and Gao, L. (2020). Oral biofilm elimination by combining iron-based nanozymes and hydrogen peroxide-producing bacteria. *Biomaterials Science*, *8*(9), 2447-2458.

[67] Wang, Y., Chen, C., Zhang, D., and Wang, J. (2020). Bifunctionalized novel Co-V MMO nanowires: Intrinsic oxidase and peroxidase like catalytic activities for antibacterial application. *Applied Catalysis B: Environmental*, *261*, 118256.

[68] Niu, J., Sun, Y., Wang, F., Zhao, C., Ren, J., and Qu, X. (2018). Photomodulated nanozyme used for a gram-selective antimicrobial. *Chemistry of Materials*, *30*(20), 7027-7033.

[69] Herget, K., P. Hubach, S. Pusch, P. Deglmann, H. Go¨tz, T. E. Gorelik, I. Y. A. Gural'skiy, F. Pfitzner, T. Link, S. Schenk, M. Pantho¨fer, V. Ksenofontov, U. Kolb, T. Opatz, R. André, W. Tremel, HaloperoXidase Mimicry by CeO2 Nanorods Combats Biofouling, *Adv. Mater.* 29 (2017) 1603823.

[70] Li, C., Sun, Y., Li, X., Fan, S., Liu, Y., Jiang, X., and Yin, J. J. (2019). Bactericidal effects and accelerated wound healing using Tb 4 O 7 nanoparticles with intrinsic oxidase-like activity. *Journal of Nanobiotechnology*, *17*(1), 1-10.

[71] Natalio, F., André, R., Hartog, A. F., Stoll, B., Jochum, K. P., Wever, R., and Tremel, W. (2012). Vanadium pentoxide nanoparticles mimic vanadium haloperoxidases and thwart biofilm formation. *Nature Nanotechnology*, *7*(8), 530-535.

[72] R. Cao-Mila'n, S. Gopalakrishnan, L. D. He, R. Huang, L. S. Wang, L. Castellanos, D. C. Luther, R. F. Landis, J. M. V. Makabenta, C. H. Li, X. Zhang, F. Scaletti, R. W. Vachet, V. M. Rotello, Thermally Gated Bio-orthogonal Nanozymes with Supramolecularly Confined Porphyrin Catalysts for Antimicrobial Uses, *Chem* 6 (2020) 1113–1124.

[73] Chen, Z., Ji, H., Liu, C., Bing, W., Wang, Z., and Qu, X. (2016). A multinuclear metal complex based DNase-mimetic artificial enzyme: matrix cleavage for combating bacterial biofilms. *Angewandte Chemie*, *128*(36), 10890-10894.

[74] Xi, J., Wei, G., Wu, Q., Xu, Z., Liu, Y., Han, J., and Gao, L. (2019). Light-enhanced sponge-like carbon nanozyme used for synergetic antibacterial therapy. *Biomaterials Science*, *7*(10), 4131-4141.

[75] Hussain, S. M., Braydich-Stolle, L. K., Schrand, A. M., Murdock, R. C., Yu, K. O., Mattie, D. M., and Terrones, M. (2009). Toxicity evaluation for safe use of

nanomaterials: recent achievements and technical challenges. *Advanced Materials*, *21*(16): 1549-1559.

[76] Zhang, J., Zou, Z., Wang, B., Xu, G., Wu, Q., Zhang, Y., and Yu, C. (2018). Lysosomal deposition of copper oxide nanoparticles triggers HUVEC cells death. *Biomaterials*, *161*: 228-239.

[77] Ginzburg, A. L., Truong, L., Tanguay, R. L., and Hutchison, J. E. (2018). Synergistic toxicity produced by mixtures of biocompatible gold nanoparticles and widely used surfactants. *ACS Nano*, *12*(6): 5312-5322.

[78] Sun, H., Zhou, Y., Ren, J., and Qu, X. (2018). Carbon nanozymes: enzymatic properties, catalytic mechanism, and applications. *Angewandte Chemie International Edition*, *57*(30), 9224-9237.

[79] Li, Y., Yuan, H., von Dem Bussche, A., Creighton, M., Hurt, R. H., Kane, A. B., and Gao, H. (2013). Graphene microsheets enter cells through spontaneous membrane penetration at edge asperities and corner sites. *Proceedings of the National Academy of Sciences*, *110*(30): 12295-12300.

[80] Zhao, Y., Ye, C., Liu, W., Chen, R., and Jiang, X. (2014). Tuning the composition of AuPt bimetallic nanoparticles for antibacterial application. *AngewandteChemie International Edition*, *53*(31): 8127-8131.

[81] Liang, M., and Yan, X. (2019). Nanozymes: from new concepts, mechanisms, and standards to applications. *Accounts of Chemical Research*, *52*(8): 2190-2200.

[82] Meng, H., Leong, W., Leong, K. W., Chen, C., and Zhao, Y. (2018). Walking the line: The fate of nanomaterials at biological barriers. *Biomaterials*, *174*, 41-53.

[83] Zhang, C., Yan, L., Wang, X., Zhu, S., Chen, C., Gu, Z., and Zhao, Y. (2020). Progress, challenges, and future of nanomedicine. *Nano Today*, *35*, 101008.

[84] Mei, L., Zhang, X., Yin, W., Dong, X., Guo, Z., Fu, W., and Zhao, Y. (2019). Translocation, biotransformation-related degradation, and toxicity assessment of polyvinylpyrrolidone-modified 2H-phase nano-MoS 2. *Nanoscale*, *11*(11): 4767-4780.

[85] Zhu, S., Li, L., Gu, Z., Chen, C., and Zhao, Y. (2020). 15 years of small: Research trends in nanosafety. *Small*, *16*(36): 2000980.

Chapter 7

The Role of Metallic Nanoparticles in Maximizing Crop Production

Divya Yadav[1], Santosh Bhukal[1,*] and Shafila Bansal[2,†]

[1] Guru Jambheshwar University of Science and Technology, Hisar, India

[2] Mehr Chand Mahajan DAV College for Women, Chandigarh, India

Abstract

For the past few decades, the biggest problem, the world is facing is population explosion which leads to sub-issues including poverty, shortage of food, and hunger. Enough crop production is a major challenge for the agricultural sector. As a result, farmers are opting for excessive use of chemicals as fertilizers and pesticides, where these chemicals make their way into water bodies through soil runoff resulting in a plethora of environmental problems including bioaccumulation and biomagnification. Thus, there is a dire need to develop green alternatives for providing nutrients to the crops to gain higher agricultural productivity without affecting the soil's health.

Among all the existing solutions, nanotechnology opens up many doors for sustainable agriculture that strengthens high-tech crop fields. Nano-sized metal oxide particles of zinc, iron, and copper have the potential to act as a carrier for nutrient delivery to the specific site in plants without any wastage. Because of their catalytic properties, nanoparticles (NPs) mimic the natural plant enzymes, which have the capability to fight against environmental stress via eliminating reactive oxygen species and augmentation of intrinsic plant functions like photosynthesis. Hence these NPs are called nanozymes.

* Corresponding Author's Email: santosh25@gjust.org.

† Corresponding Author's Email: shafibansal@yahoo.co.in.

In: Emerging Environmental Applications of Nanozymes

Editors: Seema Nara and Smriti Singh

ISBN: 979-8-88697-552-9

Additionally, nanotechnology-based pesticides have developed solutions with greater efficiency and lower toxicity by controlled release of the active ingredients. Herein, we have attempted to compile the applications of nanoparticles for maximum crop yield by acting as a carrier for the site-specific nutrient delivery, by reducing the undesirable effect of oxidative stress as well as by strengthening the endogenous defense system of plants.

Keywords: nanoparticles, nano-fertilizer, nanozymes, pesticide, stress tolerance

1. Introduction

According to an estimate by the United Nations, the world's human population will reach 9700 million by 2050, and more than 11000 million by the end of this century (Liu et al. 2021), as a result the demand for agricultural products is increasing to fulfil the appetite of growing population. Also, global agricultural crop consumption is evaluated to be 60–110% more by the year 2050 than it was in the year 2005 (Giraldo et al. 2019; Tilman et al. 2011). Hence, the rising food consumption is putting strain on agricultural lands as farmers need to grow more crops by using excessive chemicals like fertilizers, and pesticides to fulfill the demands of the burgeoning population. In developing countries like India, usage of NPK-based chemical fertilizers has severely hiked since the green revolution (Chand and Pavithra 2015). In China, pesticide use has increased rapidly in recent years, i.e., four times more than the global average pesticide use (Li et al. 2014; Zhang et al. 2015). Because of this widespread use of synthetic organic compounds to obtain maximum crop yield, the non-point source of pollution is becoming an intensifying problem. Owing to the detrimental impacts of synthetic chemicals on environmental and human health (Bindraban et al. 2020), most of them are either banned or deemed to be unsafe for human and animal health, so, the demand for eco-friendly alternatives is growing day by day (Lai et al. 2017).

Nanotechnology opens up many doors for sustainable agriculture that strengthens high-tech crop fields. It is a branch of science dealing with atoms or molecules on the nanoscale and has a significant role in the advancement of plant science (Astefanei et al. 2015; Sigmund et al. 2006; Yu et al. 2020). As per European Commission, nanomaterial (NM) is a natural or manufactured material comprising free, aggregated, or agglomerated particles, where 50% or more of them in number size distribution are in the range of 1–100 nm size (Ali et al. 2021). Among NMs, nanoparticles (NPs) are often considered as particles having a size between 1-100 nm in diameter that demonstrate unique size-dependent characteristics in comparison to the bulk material of the same element.

There are mainly two methods by which NPs can be synthesized: bottom-up and top-down approaches. In general, the bottom-up approach comprises the formation of NPs from small entities such as atoms and molecules, or by the aggregation of atoms into new nuclei, which then develop into a particle with dimensions of nano range with the assistance of different chemical and biological mechanisms. On the other hand, the top-down approach involves the synthesis of NPs from acceptable bulk material to tiny units either by using physical forces (such as crushing, spitting, and milling) or chemical reducing agents (Ali et al. 2017; Ali et al. 2021b; 2021c). NPs have a high surface-to-volume ratio compared to bulk materials and distinct physio-chemical, optical, and magnetic properties. For NPs, the size is inversely proportional to the surface area, chemical characteristics, and reactivity (Strambeanu et al. 2015). As the size goes down, the above-mentioned parameters are enhanced. For instance, gold NPs show different colors at different scales; at the macro scale, their color is yellow, whereas, at the nanoscale, their solution appears red. The change in particle size can also cause an instant change in the electrical properties of NPs (Biswal et al. 2012). These unique properties make their role significant in today's world. Nanotechnology could be beneficial in crop production as well, where, numerous studies witness its role in increasing the efficacy of inputs for farmers and reducing the loss of nutrients in an environment (Fatima et al. 2021; Guo et al. 2018; Liu & Lal 2015). For "sustainable intensification," nanotechnology has the ability to identify, protect and monitor plant illness as well as development, to boost the food quality and supply along with chemical waste minimization (Karunakaran et al. 2016; Prasad et al. 2017). Recently in plant science, NPs are getting recognition due to their remarkable positive effects on seed germination, photosynthetic activities, target-specific delivery, antimicrobial properties, and root, shoot system of plants (Jasim et al. 2017; Yang et at. 2020; Fraceto et al. 2016; Wang et al. 2016; Cai et al. 2019).

Any product that is made of NPs or uses nanotechnology to improve nutrient efficiency in the form of fertilizer is known as a nano-fertilizer (Ali et al. 2021b). Nano-fertilizers can transform agriculture in many ways; by stimulating crop growth, improving crop quality, reducing the cost of production, making stress-tolerant crops, supplying balanced nutrients, and regulating their migration to the environment. They also are capable of improving water holding capacity, soil, and microbial activity along with eco-friendly and precision farming (Chhipa et al. 2016; Duhan et al.2017). Several reports have shown the applicability of NPs as carriers for delivery and easy availability of nutrients to the specific plant site for better growth of plants (Fatima et al. 2021; Ghormade et al. 2011; Karunakaran et al. 2016).

The contribution of NPs as nanozymes, nano-fertilizer, nano-pesticides, and stress tolerance in maximum crop production is discussed in this chapter. Efforts have been made to assemble the reports depicting the positive contribution of NPs in plant growth, protection, and environmental stress tolerance.

2. Nanoparticles as Nanozymes

NPs used for the replacement of chemical fertilizer may act as nanozyme. Nanozymes are nanomaterials with inherent enzyme-like properties that have expanded in popularity in recent years due to their potential to address normal enzyme restrictions such as limited stability, high cost, and hard storage (Liang et al. 2019). Since the discovery of Fe_3O_4 NPs as peroxidase mimics in 2007, several researchers are exploring the properties of NMs as artificial enzymes (Palmqvist et al. 2017; Huang et al. 2019; Singh et al. 2021). Nanozyme has gained the attention of researchers due to its two-fold nature of nano properties and catalytic capability (Zhu et al. 2021). Particularly, metal oxides NPs such as oxides of iron, zinc, and copper have the ability to function as a carrier for the nutrition dubbed nanozymes, because of their catalytic characteristics, which have the capacity to defend against environmental strain by foraging reactive oxygen species (ROS).

The early research on nanozymes was more descriptive of their properties, however, recently structure-activity connection studies have arisen as a way to better understand the processes and influence of nanozymes. A number of studies have endeavored to systematically summarize the classes along with characteristics of nanozymes (Wei and Wang, 2013; Wu et al. 2019), their applications in environmental management, such as the recognition of polluting chemicals or entities in soil and water, the devastation of multi-drug resilient bacteria, and degradation of carbon-based pollutants from wastewater (Meng et al. 2020). Nanozymes with antioxidant activity have recently been discovered to regulate ROS in plant defense responses to environmental stressors. Among all NPs, metal oxide nanozymes demonstrate tremendous benefits in this regard (Lu et al. 2020; Zhao et al. 2020; Liu et al. 2019; Wu et al. 2017).

Enzymology has developed many methodologies, and numerous enzymes have been thoroughly investigated, all of which have influenced the structure and development of nanozymes and will continue to do so (Wang et. at. 2020). Substrate-binding and catalytically active sites on nanozymes mimicking peroxidase have been found efficient in particularly disabling the carboxyl, ketonic carbonyl, or hydroxyl groups to regulate endogenous antioxidant defense and counter the undesirable oxidative stress (Sun et al. 2015). The activity of nanozymes can also be predicted using theoretical catalytic models. For example, to forecast the enzyme-like activities of different nanomaterials, a density functional theory-based method was developed (Shen et al. 2020), revealing catalytic activity in the cone-shaped plot as a purpose of basic energy-based descriptors. Abundant nanozymes have been designed to structurally replicate regular enzymes and control their activity or selectivity, with many being inspired by real enzymes (Wei et al. 2021).

2.1. Pathways of Intake

NPs can enter into the plants in two ways, from leaf to root (foliar spray), and root to leaf (soil drenching). Out of the two applying modes, foliar application proved to be more effective in both NPs transport and delivery in comparison to soil drenching (Su et al. 2019). Plant-NP interactions are generally divided into three stages: NPs deposition on the surfaces of plants (including its root, stem, and leaf), penetration of NPs through the epidermis and cuticle of the plant, and their transport as well as transformation inside the plant. While all leaves have the same basic structure (having an epidermis with stomata, vascular tissue, and mesophyll) however, environmental conditions like availability of water, the intensity of light, biological niches, temperature, and various stressors influence how these components are arranged in each plant.

NPs may be absorbed by the plants via the above-mentioned routes and this absorption in plants varies depending on the species as well as growing circumstances. Particle movement via pores is regulated by the cell wall of plant cells (Chen et al. 2010; Kurepa et al. 2010). When NPs are taken up through the cell, they break the cell wall and are often carried up. In a number of studies, NPs are recognized to induce wider holes in the cell wall by which larger NPs may enter (Kurepa et al. 2010; Li et al. 2014; Servin et al. 2012).

Whereas, within cells, NPs are carried by apoplastic or symplastic routes, and they are transferred between cells via plasmodesmata (Rico et al. 2011). Apoplastic transport happens across cell walls and intercellular air gaps (Sattelmacher 2001; Wang et al. 2012a) and the symplastic route transports water and other chemicals across nearby cells cytoplasm (Roberts and Oparka 2003). NPs enter the root cylinders and then use the apoplastic route to reach the plant's aerial portions (Sun et al. 2014). The type and nature of NPs, as well as the plant species, have a big impact on translocation in plant tissues. To illustrate, in soybean plants, CeO NPs translocated through the xylem from roots to shoots, whilst Zn is bio-transformed into Zn citrate (Hernandez-Viezcas et al. 2013).

3. Nano-Fertilizers

Nano-fertilizers are nutritional fertilizers that are made up of nanostructured formulations that may be delivered to plants and allows for effective absorption or gradual release of active components, much like traditional fertilizers (Sachan 2022). Conventional fertilizers are mostly applied to the soil by farmers or producers by various methods including surface broadcasting, subsurface application, fertigation, or irrigation, where, a considerable amount of them is discharged into the atmosphere or enters the water bodies, and hence degrade our ecosystems. For example, 75% of the nitrogen in urea is lost by thermal decomposition (as ammonia, or emission as nitrous oxide or nitric oxide) and nitrogen trioxide seeps into water bodies after application in the field. As a result, present nitrogen fertilizers have a low nitrogen capacity utilization (about 20%), resulting in eutrophication and a rise in greenhouse gas emissions. Among the vital macronutrients, 40–70% of applied nitrogen, 80–90% of phosphorous, and 50–90% of potassium tend to get lost, and the use of NMs as agrochemicals is being thoroughly investigated to reduce the loss of these vital macronutrients (Chand and Pavithra 2015).

An application of Ag NPs (concentration ranging from 20 -100 ppm) as fertilizer has shown a positive effect on the shoot and root length, the surface area of the leaf, protein, and chlorophyll content in plants like a common bean (*P. vulgaris L.)* and maize (*Z. mays L.*). The findings have revealed that even low concentrations of Ag NPs have a constructive influence on plant development (Zea and Salama 2012). Like Ag, Zn is also a crucial element for plants (Hafeez et al. 2013), and it is taken mostly as Zn^{2+}. Many enzymes use this vitamin as metal components, cofactors, and other governing variables, making it particularly important for plant physiological responses (Fatima et al. 2021). The influence of ZnO NPs on plant development and its use in agriculture is widely explored by numerous researchers, (Prasad et al. 2014; Jamdagni et al. 2018). The use of ZnO NPs in the rice crop at varied doses may result in greater rice production with larger panicle number (4.83–13.14 percent), spikelet per panicle (4.81–10.69 percent), grain weight (3.82–6.62 percent), and high filled grain rate (0.28–2.36 percent) (Zhang et al. 2021). In cucumber fruits, Mg content can be boosted by 400 mg kg^{-1} by using ZnO NPs. (Zhao et al. 2014). Maize crop shows encouraging results when very lower quantities of 400 mg Zn L^{-1} as NPs are applied as compared to 2000 mg Zn L^{-1} as $ZnSO_4$, where, plant height, leaf area, dry weight, grain production, cob length, and the number of grains per row all increased (Šebesta et al. 2021). In comparison to the application of 2000 mg Zn L^{-1} as chelated Zn, the foliar spray of ZnO NPs of 133 mg Zn L^{-1} demonstrates better performance in peanut plants. (Mielcarz et al. 2021).

The micronutrient, Iron, is also required for the growth and development of plants. Cell metabolism, photosynthesis, and respiration are only a few of the physiological and biochemical events in which it is involved. For example, it is required for the production of certain chlorophyll-protein complexes in chloroplasts, and its shortage results in the yellowing of leaves and a reduction in photosynthetic capability. It is also a co-factor for several enzymes, and its activity is tightly linked to it. At neutral pH levels, iron normally forms insoluble ferric complexes, making it inaccessible to plants. As a result, Fe-enriched fertilizers may be able to help plants to get the required amount of Fe. In comparison to control and/or synthetic Fe sources, Fe nano-fertilizers have boosted germination and improved crop growth in many experiments (Rui et al., 2016). Optimized use of Fe_2O_3 NPs in the maize crop may greatly enhance root elongation (Elanchezhian et al., 2017). For spinach plants, stem and root length can be greatly promoted by the absorption of Fe_2O_3 NPs. Spinach growth (*Spinacia oleracea L.*) may even be improved by using FeS_2 NPs (Jeyasubramanian et al. 2016). Studies have also shown a faster growth of Fe NPs-treated peanut (*Arachis hypogaea L.*) plants than the non-treated ones (Rui et al, 2016). Green gram (Vigna radiate L.) has to show a longer radical length during germination and greater fresh biomass on Fe NPs treatment (Samrot et al. 2020). Titanium is another micronutrient required by plants for their development. The use of Ti as TiO_2NP on Canola seeds has shown improved seedling vigor and seed germination (Mahmoodzadeh et al. 2015). The stimulation of seed germination in wheat has also been reported on the application of TiO_2 NP (Feizi et al. 2012).

The growing trend shows that developing and using new forms of fertilizers is one of the only realistic alternatives for feeding a predicted world population of 9.6 billion by 2050 or more without putting ecosystems and the environment in significant jeopardy. As a result, identifying and implementing accessible breakthrough technologies in fertilizer research and development is critical.

4. Nano-Pesticides

Conventional pesticides can eradicate exotic species including aphids, grasshoppers, noctuid, and moths of nectarine but they concurrently destroy other useful insects like bees. The mortality of dangerous pests causes a reduction in the fertility of the soil and the quality of food. It's worth noting that just 0.1 percent of used pesticides hit the target species (Hu and Xianyu 2021). Residual pesticides have the potential to harm the ecosystem, including topsoil and groundwater. Because chemical pesticides may accumulate through food chains, they may eventually represent serious harm to humans. Nanotechnology-based enhanced pesticides (nano-pesticides) have developed solutions with greater efficiency and lower toxicity. From the formulated nano-pesticide, the active ingredients are released in a controlled manner which has the potential to outspread the contact of target organisms, and thus reduce the quantity of vital dose for pest management (Roy et al. 2014; Guan et al. 2018a, b; Liang et al. 2018).

In agriculture, a variety of nanomaterials have been employed as nano-pesticides, including NPs of copper and copper oxide, silica, silver, sulfur, zinc oxide, chitosan NPs, titanium dioxide, and reduced graphene nanosheets (Chen et al. 2016; Shanker et al. 2018; Singh et al. 2018; Jaffri et al. 2019; Chen et al. 2019; Prasad et al. 2019; Paralikar et al. 2019). When nano-pesticides are exposed to UV radiation, the reactive oxygen species such as hydroxyl radicals ($^{\bullet}OH$), superoxide anions (O^{2-}), and hydrogen peroxide (H_2O_2) are generated, which results in a strong plant defense system because of their antimicrobial properties. NPs can also be applied as nanocarriers for traditionally used pesticides and can also track the presence of pesticides accurately by mounting with organic dyes (Hu and Xianyu 2021). Many studies have exposed the potential of NPs to act as nano-pesticides (Buteler et al. 2015; Selvan et al. 2018; Ayoub et al. 2018). For instance, *Helicoverpa armigera* (Lepidoptera: Noctuidae) is very sensitive to Ag NPs, particularly, in cotton leaves, all the larval instars and pupae of *Helicoverpa armigera* died quickly when exposed to 10 ppm of Ag NPs (Durga et al. 2014).

CuO NPs have been recommended for regulating distinct instar larvae of the cotton leaf worm, *Spodoptera littoralis* (Ayoub et al. 2018). These NPs have also proved their capability in enhancing the mortality rate of Aedes aegypti larvae (Selvan et al. 2018). Apart from silver and copper NPs, nanostructured alumina has also shown their pesticidal effectiveness against pests, mainly found on wheat crops like *Sitophilus oryzae* and *Rhizopertha dominica*, where, an increased death rate has been observed on continuous exposure of nanostructured to these crops and 100 percent of *R. dominica* and *S. oryzae* adults has died in three days on exposure of 250 ppm of nanostructured alumina (Al_2O_3) (Buteler et al. 2015).

5. Nanoparticles for Strengthening Stress Tolerance Power

Various abiotic stressors viz. drought, floods, heat, hail, salt, heavy metals, and mineral shortages restrict crop output by negatively affecting plant development and productivity, which represent a serious threat to global food security. Drought is a severe abiotic factor that reduces agricultural output dramatically. The ever-increasing problem of water shortage is now disturbing agricultural yield by reducing green belts throughout the planet. In addition to the development of drought-resistant crops, the usage of stress-ameliorative ingredients such as nano-fertilizers has the potential to significantly reduce the harmful effects of drought on plants (Seleiman et al. 2018; 2021).

Under the worst environmental conditions like salinity, flooding, drought, excessive light, and heat stress, nanomaterials have shown promising results in plant growth. The SiO_2 NPs application on the pumpkin under salinity conditions shows enhanced growth, reduced chlorophyll degradation, and improved antioxidant enzyme activity (Ashkavand et al. 2018). Furthermore, the evaluation of salinity conditions using Ag nanoparticles on tomato negatively affect its height, and traits of fruit (Almutairi et al. 2016).

The evaluation of the role of nano-Ca, nano-mono potassium phosphate, and Nano-urea modified with hydroxyapatite in crops viz. tomato and almond, under salinity stress conditions, shows a significant increase in germination rate, length, number of flowers, and diameter of the plant (Tantawy et al. 2015; Alharby et al. 2017; Badran et al. 2018). Positive impacts like increased antioxidant enzymes and decreased accumulation of reactive oxygen species on spinach plants even in stressed conditions of excessive light can be noticed by using TiO_2 NPs (Hong et al. 2005). For tomato plants, under heat stress conditions, TiO_2 NPs can result in the cooling of leaves, enhanced photosynthesis, and regulated energy dissipation (Qi et al. 2013).

An induced chromosomal aberration may be observed under oxidative stress conditions in an onion crop when it is exposed to CuO, Al_2O_3, and TiO_2 NPs (Ahmed et al. 2018). Multi-walled carbon nanotubes (MWCNT) also illustrate an increased water uptake, plant growth, and assimilation of CO_2 for cabbage and Zucchini crops under the stressed environment of drought and salinity (Amooaghaie et al. 2018, Van Eerd et al. 2017).

6. Challenges and Future Aspects

Nanozymes' excellent stability and catalytic activity make it simple to make catalysts and medicines for a broader range of uses. However, nanozyme selectivity and activity must be improved, and biocompatibility is a difficult obstacle to overcome (Huang et al. 2009; Wang et al. 2019). Aggregation is a critical challenge to address throughout the synthesis process since it reduces available surface area, which impacts their reactivity. Furthermore, when aggregate sizes get larger, their mobility in porous media decreases, affecting NMs responsiveness and behavior (Lowry et al., 2012).

Soil organic matter (SOM) is another important component that has a significant impact on the fate and behavior of NMs, mostly through adsorption and stability (Lei et al., 2018). Adsorption of SOM would result in the surface covering of NMs, reducing their active surface area and so mitigating the potential impacts of NMs (Li et al., 2008). Soil microorganisms can be directly impacted by the toxicity of additional NMs or indirectly by magnifying the bioavailability of other harmful substances already present in soil (Simonet and Valcárcel 2009).

The toxicity of these NPs on microbial activity and functioning varies greatly depending on the kind of NPs, with inorganic NPs (metal and metal oxide) being more hazardous than organic NPs (fullerenes and carbon nanotubes) (Rajput et al., 2018). So, to overcome these problems, it is important to explore more in the field of nano-fertilizers and nanozymes. To select the appropriate metal oxide NPs more in-depth research studies are required.

Conclusion

To conclude, the growing population necessitates the production of more crops. Researchers are attempting to find innovative ways to boost food production without putting more strain on natural resources. Nanotechnology is proving to be the best way to tackle the issue of employing chemicals to boost output. The use of nanomaterials as nano-fertilizers and nano-pesticides can be an effective and environmentally friendly way to stimulate plant development. Metal oxide nanoparticles of Fe and Zn can function as nanosized micronutrients for plants, while Cu and Ag can provide the greatest pest management outcomes. Along with nano-fertilizers, metal NPs, such as Fe, can act as nanozymes, enhancing catalytic capabilities to speed up plants' reactions and providing strength to withstand harsh situations.

There is now a need to construct nanoscale macro and micronutrient fertilizers, as well as better synthesis and dosage optimization. For commercial availability of nano agro-materials, it is requisite to bridge the gap between laboratories and farmers.

References

Ahmed, Bilal, Mohammad Shahid, Mohammad Saghir Khan, and Javed Musarrat. "Chromosomal aberrations, cell suppression, and oxidative stress generation induced by metal oxide nanoparticles in onion (Allium cepa) bulb." *Metallomics* 10, no. 9 (2018): 1315-1327.

Alharby, Hesham F., Ehab MR Metwali, Michael P. Fuller, and Amal Y. Aldhebiani. "Impact of application of zinc oxide nanoparticles on callus induction, plant regeneration, element content and antioxidant enzyme activity in tomato (Solanum lycopersicum Mill) under salt stress." *Archives of Biological Sciences* 68(4):723-735 (2016).

Ali, Sameh S., Michael Kornaros, Alessandro Manni, Rania Al-Tohamy, Abd El-Raheem R. El-Shanshoury, Ibrahim M. Matter, Tamer Elsamahy, Mabrouk Sobhy, and Jianzhong Sun. "Advances in microorganisms-based biofertilizers: Major mechanisms and applications." In *Biofertilizers*, pp. 371-385. Woodhead Publishing, 2021.

Ali, Sameh S., Osama M. Darwesh, Michael Kornaros, Rania Al-Tohamy, Alessandro Manni, Abd El-Raheem R. El-Shanshoury, Metwally A. Metwally, Tamer Elsamahy, and Jianzhong Sun. "Nano-biofertilizers: Synthesis, advantages, and applications." In *Biofertilizers*, pp. 359-370. Woodhead Publishing, 2021.

Ali, Sameh S., Rania Al-Tohamy, Eleni Koutra, Mohamed S. Moawad, Michael Kornaros, Ahmed M. Mustafa, Yehia A. G. Mahmoud, Badr, A., Osman, M. E. H., Elsamahy,

T., Jiao, H., & Sun, J. "Nanobiotechnological advancements in agriculture and food industry: Applications, nanotoxicity, and future perspectives." *Science of the Total Environment* 792 (2021): 148359.

Ali, Shahid, Ibrahim Khan, Safyan Akram Khan, Manzar Sohail, Riaz Ahmed, Ateeq ur Rehman, Muhammad Shahid Ansari, and Mohamed Ali Morsy. "Electrocatalytic performance of Ni@ Pt core–shell nanoparticles supported on carbon nanotubes for methanol oxidation reaction." *Journal of Electroanalytical Chemistry* 795 (2017): 17-25.

Almutairi, Zainab M. "Influence of silver nano-particles on the salt resistance of tomato (Solanum lycopersicum) during germination." *Int. J. Agric. Biol* 18, no. 2 (2016): 449-457.

Amooaghaie, Rayhaneh, Fatemeh Tabatabaei, and Alimohammad Ahadi. "Alterations in HO-1 expression, heme oxygenase activity and endogenous NO homeostasis modulate antioxidant responses of Brassica nigra against nano silver toxicity." *Journal of Plant Physiology* 228 (2018): 75-84.

Ashkavand, Peyman, Mehrdad Zarafshar, Masoud Tabari, Javad Mirzaie, Amirreza Nikpour, Seyed Kazem Bordbar, Daniel Struve, and Gustavo Gabriel Striker. "Application of SiO2 nanoparticles as pretreatment alleviates the impact of drought on the physiological performance of Prunus mahaleb (Rosaceae)." *Boletín de la Sociedad Argentina de Botánica* 53, no. 2 (2018): 207-219.

Astefanei, Alina, Oscar Núñez, and Maria Teresa Galceran. "Characterisation and determination of fullerenes: a critical review." *Analytica Chimica Acta* 882 (2015): 1-21.

Ayoub, Haytham A., Mohamed Khairy, Salaheldeen Elsaid, Farouk A. Rashwan, and Hanan F. Abdel-Hafez. "Pesticidal activity of nanostructured metal oxides for generation of alternative pesticide formulations." *Journal of Agricultural and Food Chemistry* 66, no. 22 (2018): 5491-5498.

Badran, Antar, and Igor Savin. "Effect of nano-fertilizer on seed germination and first stages of bitter almond seedlings' growth under saline conditions." *BioNanoScience* 8, no. 3 (2018): 742-751

Biswal, Susanta Kumar, Ashok Kumar Nayak, Umesh Kumar Parida, and P. L. Nayak. "Applications of nanotechnology in agriculture and food sciences." *Int J Sci Innov Discov* 2, no. 1 (2012): 21-36.

Buteler, Micaela, S. W. Sofie, D. K. Weaver, D. Driscoll, J. Muretta, and Teodoro Stadler. "Development of nanoalumina dust as insecticide against Sitophilus oryzae and Rhyzopertha dominica." *International Journal of Pest Management* 61, no. 1 (2015): 80-89.

Cai, Lin, Changyun Liu, Guangjin Fan, Chaolong Liu, and Xianchao Sun. "Preventing viral disease by ZnONPs through directly deactivating TMV and activating plant immunity in Nicotiana benthamiana." *Environmental Science: Nano 6,* no. 12 (2019): 3653-3669.

Chen, Juanni, Shili Li, Jinxiang Luo, Rongsheng Wang, and Wei Ding. "Enhancement of the antibacterial activity of silver nanoparticles against phytopathogenic bacterium Ralstonia solanacearum by stabilization." *Journal of Nanomaterials* 2016 (2016).

Chen, Juanni, Shuyu Mao, Zhifeng Xu, and Wei Ding. "Various antibacterial mechanisms of biosynthesized copper oxide nanoparticles against soilborne Ralstonia solanacearum." *RSC Advances* 9, no. 7 (2019): 3788-3799.

Chen, Ran, Tatsiana A. Ratnikova, Matthew B. Stone, Sijie Lin, Mercy Lard, George Huang, JoAn S. Hudson, and Pu Chun Ke. "Differential uptake of carbon nanoparticles by plant and mammalian cells." *Small* 6, no. 5 (2010): 612-617.

Chhipa, Hemraj, and Piyush Joshi. "Nanofertilisers, nanopesticides and nanosensors in agriculture." In *Nanoscience in Food and Agriculture* 1, pp. 247-282. Springer, Cham, 2016.

Devi, G. Durga, K. Murugan, and C. Panneer Selvam. "Green synthesis of silver nanoparticles using Euphorbia hirta (Euphorbiaceae) leaf extract against crop pest of cotton bollworm, Helicoverpa armigera (Lepidoptera: Noctuidae)." *Journal of Biopesticides* 7 (2014): 54.

Duhan, Joginder Singh, Ravinder Kumar, Naresh Kumar, Pawan Kaur, Kiran Nehra, and Surekha Duhan. "Nanotechnology: The new perspective in precision agriculture." *Biotechnology Reports* 15 (2017): 11-23.

Elanchezhian, Rajamanickam, Dameshwar Kumar, Kulasekaran Ramesh, Ashish Kumar Biswas, Arti Guhey, and Ashok Kumar Patra. "Morpho-physiological and biochemical response of maize (Zea mays L.) plants fertilized with nano-iron (Fe_3O_4) micronutrient." *Journal of Plant Nutrition* 40, no. 14 (2017): 1969-1977.

Fatima, Faria, Arshya Hashim, and Sumaiya Anees. "Efficacy of nanoparticles as nanofertilizer production: a review." *Environmental Science and Pollution Research* 28, no. 2 (2021): 1292-1303.

Ghormade, Vandana, Mukund V. Deshpande, and Kishore M. Paknikar. "Perspectives for nano-biotechnology enabled protection and nutrition of plants." *Biotechnology Advances* 29, no. 6 (2011): 792-803.

Giraldo, Juan Pablo, Honghong Wu, Gregory Michael Newkirk, and Sebastian Kruss. "Nanobiotechnology approaches for engineering smart plant sensors." *Nature Nanotechnology* 14, no. 6 (2019): 541-553.

Hernandez-Viezcas, Jose A., Hiram Castillo-Michel, Joy Cooke Andrews, Marine Cotte, Cyren Rico, Jose R. Peralta-Videa, Yuan Ge, John H. Priester, Patricia Ann Holden, and Jorge L. Gardea-Torresdey. "*In situ* synchrotron X-ray fluorescence mapping and speciation of CeO_2 and ZnO nanoparticles in soil cultivated soybean (Glycine max)." *ACS Nano* 7, no. 2 (2013): 1415-1423.

Hong, F. S., P. Yang, F. Q. Gao, Ch Liu, L. Zheng, F. Yang, and J. Zhou. "Effect of nano-TiO_2 on spectral characterization of photosystem II particles from spinach." *Chem Res Chin Univ 21*, no. 2 (2005): 196-200.

Hu, Jing, and Yunlei Xianyu. "When nano meets plants: A review on the interplay between nanoparticles and plants." *Nano Today* 38 (2021): 101143.

Jaffri, Shaan Bibi, and Khuram Shahzad Ahmad. "Phytofunctionalized silver nanoparticles: green biomaterial for biomedical and environmental applications." *Reviews in Inorganic Chemistry* 38, no. 3 (2018): 127-149.

Jasim, B., Roshmi Thomas, Jyothis Mathew, and E. K. Radhakrishnan. "Plant growth and diosgenin enhancement effect of silver nanoparticles in Fenugreek (Trigonella foenum-graecum L.)." *Saudi Pharmaceutical Journal* 25, no. 3 (2017): 443-447.

Kurepa, Jasmina, Tatjana Paunesku, Stefan Vogt, Hans Arora, Bryan M. Rabatic, Jinju Lu, M. Beau Wanzer, Gayle E. Woloschak, and Jan A. Smalle. "Uptake and distribution of ultrasmall anatase TiO_2 Alizarin red S nanoconjugates in Arabidopsis thaliana." *Nano Letters* 10, no. 7 (2010): 2296-2302.

Lei, Cheng, Yuqing Sun, Daniel CW Tsang, and Daohui Lin. "Environmental transformations and ecological effects of iron-based nanoparticles." *Environmental Pollution* 232 (2018): 10-30.

Li, Dong, Delina Y. Lyon, Qilin Li, and Pedro JJ Alvarez. "Effect of soil sorption and aquatic natural organic matter on the antibacterial activity of a fullerene water suspension." *Environmental Toxicology and Chemistry: An International Journal* 27, no. 9 (2008): 1888-1894.

Liu, Ruiqiang, and Rattan Lal. "Potentials of engineered nanoparticles as fertilizers for increasing agronomic productions." *Science of the Total Environment* 514 (2015): 131-139.

Liu, Yinglin, Zhenggao Xiao, Feiran Chen, Le Yue, Hua Zou, Jinze Lyu, and Zhenyu Wang. "Metallic oxide nanomaterials act as antioxidant nanozymes in higher plants: Trends, meta-analysis, and prospect." *Science of the Total Environment* 780 (2021): 146578.

Lowry, Gregory V., Kelvin B. Gregory, Simon C. Apte, and Jamie R. Lead. *"Transformations of Nanomaterials in the Environment."* (2012): 6893-6899.

Mielcarz-Skalska, Lidia, Beata Smolińska, and Katarzyna Włodarczyk. "Nanoparticles as Potential Improvement for Conventional Fertilisation in the Cultivation of Raphanus sativus var. sativus." *Agriculture* 11, no. 11 (2021): 1067.

Palmqvist, NG Martin, Gulaim A. Seisenbaeva, Peter Svedlindh, and Vadim G. Kessler. "Maghemite nanoparticles acts as nanozymes, improving growth and abiotic stress tolerance in Brassica napus." *Nanoscale Research Letters* 12, no. 1 (2017): 1-9.

Paralikar, Priti, Avinash P. Ingle, Vaibhav Tiwari, Patrycja Golinska, Hanna Dahm, and Mahendra Rai. "Evaluation of antibacterial efficacy of sulfur nanoparticles alone and in combination with antibiotics against multidrug-resistant uropathogenic bacteria." *Journal of Environmental Science and Health*, Part A 54, no. 5 (2019): 381-390.

Prasad, Ashwini, Syed Baker, Nagendra Prasad, Aishwarya Tripurasundari Devi, S. Satish, Farhan Zameer, and Chandan Shivamallu. "Phytogenic synthesis of silver nanobactericides for anti-biofilm activity against human pathogen H. pylori." *SN Applied Sciences* 1, no. 4 (2019): 1-7.

Prasad, T. N. V. K. V., P. Sudhakar, Y. Sreenivasulu, P. Latha, V. Munaswamy, K. Raja Reddy, T. S. Sreeprasad, P. R. Sajanlal, and T. Pradeep. "Effect of nanoscale zinc oxide particles on the germination, growth and yield of peanut." *Journal of plant nutrition* 35, no. 6 (2012): 905-927.

Qi, Mingfang, Yufeng Liu, and Tianlai Li. "Nano-TiO_2 improve the photosynthesis of tomato leaves under mild heat stress." *Biological Trace Element Research* 156, no. 1 (2013): 323-328.

Rajput, Vishnu D., Tatiana Minkina, Svetlana Sushkova, Viktoriia Tsitsuashvili, Saglara Mandzhieva, Andrey Gorovtsov, Dina Nevidomskyaya, and Natalya Gromakova. "Effect of nanoparticles on crops and soil microbial communities." *Journal of Soils and Sediments* 18, no. 6 (2018): 2179-2187.

Rico, Cyren M., Sanghamitra Majumdar, Maria Duarte-Gardea, Jose R. Peralta-Videa, and Jorge L. Gardea-Torresdey. "Interaction of nanoparticles with edible plants and their possible implications in the food chain." *Journal of Agricultural and Food Chemistry* 59, no. 8 (2011): 3485-3498.

Roberts, Alison. "Plasmodesmata and the control of symplastic transport. Plant Cell Environ." *Plant, Cell and Environment* 26 (2003): 103-124.

Rui, Mengmeng, Chuanxin Ma, Yi Hao, Jing Guo, Yukui Rui, Xinlian Tang, Qi Zhao et al. "Iron oxide nanoparticles as a potential iron fertilizer for peanut (Arachis hypogaea)." *Frontiers in Plant Science* 7 (2016): 815.

Samrot, Antony V., C. SaiPriya, Jenifer Selvarani, Jane Cypriyana PJ, Y. Lavanya, P. Soundarya, Sherly Priyanka RB, P. Sangeetha, and Reji Joseph Varghese. "A study on influence of superparamagnetic iron oxide nanoparticles (SPIONs) on green gram (Vigna radiata L.) and earthworm (Eudrilus eugeniae L.)." *Materials Research Express* 7, no. 5 (2020): 055002.

Sattelmacher, Burkhard. "The apoplast and its significance for plant mineral nutrition." *New Phytologist* 149, no. 2 (2001): 167-192.

Seleiman, Mahmoud F., Khalid F. Almutairi, Majed Alotaibi, Ashwag Shami, Bushra Ahmed Alhammad, and Martin Leonardo Battaglia. "Nano-fertilization as an emerging fertilization technique: why can modern agriculture benefit from its use?." *Plants* 10, no. 1 (2020): 2.

Selvan, Sekaran Muthamil, Kabali Vijai Anand, Kasivelu Govindaraju, Selvaraj Tamilselvan, Vijayakumar Ganesh Kumar, Kizhaeral Sevathapandian Subramanian, Malaisamy Kannan, and Kalimuthu Raja. "Green synthesis of copper oxide nanoparticles and mosquito larvicidal activity against dengue, zika and chikungunya causing vector Aedes aegypti." *IET Nanobiotechnology* 12, no. 8 (2018): 1042-1046.

Shankar, Shiv, and Jong-Whan Rhim. "Preparation of sulfur nanoparticle-incorporated antimicrobial chitosan films." *Food Hydrocolloids* 82 (2018): 116-123.

Shen, Xiaomei, Zhenzhen Wang, Xingfa Gao, and Yuliang Zhao. "Density functional theory-based method to predict the activities of nanomaterials as peroxidase mimics." *ACS Catalysis* 10, no. 21 (2020): 12657-12665.

Sigmund, Wolfgang, Junhan Yuh, Hyun Park, Vasana Maneeratana, Georgios Pyrgiotakis, Amit Daga, Joshua Taylor, and Juan C. Nino. "Processing and structure relationships in electrospinning of ceramic fiber systems." *Journal of the American Ceramic Society* 89, no. 2 (2006): 395-407.

Simonet, Bartolomé M., and Miguel Valcárcel. "Monitoring nanoparticles in the environment." *Analytical and bioanalytical chemistry* 393, no. 1 (2009): 17-21.

Singh, Ajey, N. áB Singh, Shadma Afzal, Tanu Singh, and Imtiyaz Hussain. "Zinc oxide nanoparticles: a review of their biological synthesis, antimicrobial activity, uptake, translocation and biotransformation in plants." *Journal of Materials Science* 53, no. 1 (2018): 185-201.

Singh, Davinder, Devendra Sillu, Anil Kumar, and Shekhar Agnihotri. "Dual nanozyme characteristics of iron oxide nanoparticles alleviate salinity stress and promote the growth of an agroforestry tree, Eucalyptus tereticornis Sm." *Environmental Science: Nano* 8, no. 5 (2021): 1308-1325.

Strambeanu, Nicolae, Laurentiu Demetrovici, Dan Dragos, and Mihai Lungu. "Nanoparticles: Definition, classification and general physical properties." In *Nanoparticles' Promises and Risks*, pp. 3-8. Springer, Cham, 2015

Su, Yiming, Vanessa Ashworth, Caroline Kim, Adeyemi S. Adeleye, Philippe Rolshausen, Caroline Roper, Jason White, and David Jassby. "Delivery, uptake, fate, and transport of engineered nanoparticles in plants: a critical review and data analysis." *Environmental Science: Nano* 6, no. 8 (2019): 2311-2331.

Sun, Dequan, Hashmath I. Hussain, Zhifeng Yi, Rainer Siegele, Tom Cresswell, Lingxue Kong, and David M. Cahill. "Uptake and cellular distribution, in four plant species, of fluorescently labeled mesoporous silica nanoparticles." *Plant Cell Reports* 33, no. 8 (2014): 1389-1402.

Sun, Hanjun, Andong Zhao, Nan Gao, Kai Li, Jinsong Ren, and Xiaogang Qu. "Deciphering a nanocarbon-based artificial peroxidase: chemical identification of the catalytically active and substrate-binding sites on graphene quantum dots." *Angewandte Chemie International Edition 54*, no. 24 (2015): 7176-7180.

Tantawy, A. S., Y. A. M. Salama, M. A. El-Nemr, and A. M. R. Abdel-Mawgoud. "Nano silicon application improves salinity tolerance of sweet pepper plants." *Int J ChemTech Res* 8, no. 10 (2015): 11-17.

Tilman, David, Christian Balzer, Jason Hill, and Belinda L. Befort. "Global food demand and the sustainable intensification of agriculture." *Proceedings of the National Academy of Sciences 108*, no. 50 (2011): 20260-20264.

Van Eerd, L. L., J. J. D. Turnbull, C. J. Bakker, R. J. Vyn, A. W. McKeown, and S. M. Westerveld. "Comparing soluble to controlled-release nitrogen fertilizers: storage cabbage yield, profit margins, and N use efficiency." *Canadian Journal of Plant Science* 98, no. 4 (2017): 815-829.

Wang, Hui, Kaiwei Wan, and Xinghua Shi. "Recent advances in nanozyme research." *Advanced Materials* 31, no. 45 (2019): 1805368.

Wang, Qiang, Xingmao Ma, Wen Zhang, Haochun Pei, and Yongsheng Chen. "The impact of cerium oxide nanoparticles on tomato (Solanum lycopersicum L.) and its implications for food safety." *Metallomics* 4, no. 10 (2012): 1105-1112.

Wang, Zhuoran, Ruofei Zhang, Xiyun Yan, and Kelong Fan. "Structure and activity of nanozymes: Inspirations for de novo design of nanozymes." *Materials Today* 41 (2020): 81-119.

Wei, Hui, and Erkang Wang. "Nanomaterials with enzyme-like characteristics (nanozymes): next-generation artificial enzymes." *Chemical Society Reviews* 42, no. 14 (2013): 6060-6093.

Wei, Hui, Lizeng Gao, Kelong Fan, Juewen Liu, Jiuyang He, Xiaogang Qu, Shaojun Dong, Erkang Wang, and Xiyun Yan. "Nanozymes: A clear definition with fuzzy edges." *Nano Today* 40 (2021): 101269.

Wu, Jiangjiexing, Xiaoyu Wang, Quan Wang, Zhangping Lou, Sirong Li, Yunyao Zhu, Li Qin, and Hui Wei. "Nanomaterials with enzyme-like characteristics (nanozymes): next-generation artificial enzymes (II)." *Chemical Society Reviews* 48, no. 4 (2019): 1004-1076.

Yang, Xueling, Darioush Alidoust, and Chunyan Wang. "Effects of iron oxide nanoparticles on the mineral composition and growth of soybean (Glycine max L.) plants." *Acta Physiologiae Plantarum* 42, no. 8 (2020): 1-11.

Yu, Sujie, Jianzhong Sun, Yifei Shi, Qianqian Wang, Jian Wu, and Jun Liu. "Nanocellulose from various biomass wastes: Its preparation and potential usages towards the high value-added products." *Environmental Science and Ecotechnology* 5 (2021): 100077.

Zheng, Lei, Fashui Hong, Shipeng Lu, and Chao Liu. "Effect of nano-TiO_2 on strength of naturally aged seeds and growth of spinach." *Biological Trace Element Research* 104, no. 1 (2005): 83-91.

Zhu, Hengjia, Peng Liu, Lizhang Xu, Xin Li, Panwang Hu, Bangxiang Liu, Jianming Pan, Fu Yang, and Xiangheng Niu. "Nanozyme-Participated Biosensing of Pesticides and Cholinesterases: A Critical Review." *Biosensors* 11, no. 10 (2021): 382.

Chapter 8

Biomedical Applications of Nanozymes

Bikash Ranjan Jena and Gurudutta Pattnaik*
School of Pharmacy and Life Sciences,
Centurion University of Technology and Management,
Bhubaneswar, Odisha, India

Abstract

In recent years, researchers worldwide have been focusing on nanomaterials that move in an enzyme-like manner in scientific and organized attempts to replace enzymes in healthcare and industrial manufacturing applications. The use of nanozymes in biological research and development has recently emerged as a new platform for innovation and advancement. It is possible to see high sensitivity and rapid repeatability when utilizing nanozymes, which have been lately applied as the driving force for analytical procedures and statistical quality control.

Scientists worldwide are increasingly interested in learning more about the diverse uses of nanoenzymes in a variety of rapidly growing domains of environmental, health, and diagnostics applications. The catalytic activities of nanozymes are used for detecting and developing sensing systems for antibiotic residues, pesticides, food pollutants, etc. The upcoming generation of nanozymes that will be derived will have even better catalytic activity and more stream substrate specificity than those currently available.

In this chapter, an effort has been made to depict the development of nanozymes in biomedical applications.

Keywords: biomolecules, reproducibility, sensitivity, therapeutic, immunoassay, nanomaterial

* Corresponding Author's Email: gurudutta.pattnaik@cutm.ac.in.

1. Introduction

Due to rapid advancement and innovation in nanoscience and nanotechnology the research interests have expanded in vast areas. One of the cutting-edge research areas with promising potential for illness management is nanozymes (Wang et al., 2020; Dong et al., 2019). Even though several nanozymes have synthesized in the area of biomedicine, it is still difficult to acquire an underlying knowledge of the variables that evaluate the enzyme-like catalytic performance and selectivity of nanozymes towards substrate based on environment and intrinsic structure for their interaction (Wang et al., 2020; Dong et al., 2019 and Wei and Wang 2013). Furthermore, an understanding of the catalytic mechanism is important for the rational design of innovative metal oxide nanozymes with inherent catalytic capacities, an understanding of catalytic mechanisms is important. This technique has been broadly carried out in biomedicine as a programmable multipurpose platform (Huang, 2019).

Many imaging techniques, such as ultrasound sonography (USG), computed tomography (CT), magnetic resonance imaging (MRI), and positron emission tomography (PET), are utilised in the process of identifying and diagnosing medical conditions (Huang, 2019). These methods are non-invasive, and few of them can create high-resolution images of the organs inside the body. Contrast chemicals are typically utilised in these bioimaging technologies to distinguish between healthy and sick tissue, identifying the tissue or organ of interest. (Bery, Curtis, and Phys 2003; Pankhurst et al., 2003; Tartaj et al., 2003; Sun and Zhang, 2008). Incorporating nanozymes and nanocomposites in bioimaging techniques into health care management systems is one of the aspects.

The emergence of nanoparticles is attributable to the advancement of science and technology in Nanosystems. Nanosilver or silver nanoparticles (AgNPs) were found and proven to have distinct physiochemical and biophysical properties compared to larger AgNPs. It was discovered that AgNPs possess high thermal and electrical conductivity, catalytic activity, chemical stability, and augmented Raman scattering. Research also found that, AgNPs possess antibacterial, antifungal, antiviral, and anti-inflammatory effects (Ge et al., 2014). Nanosilver has received interest for therapeutic applications, including wound dressings, surgical equipment, targeted drug delivery and bone prosthesis due to its unique qualities and characteristics (Ge et al., 2014). Nanoceria's intriguing antioxidant enzyme like properties (super oxide dismutase (SOD), oxidase, and catalase mimetics) indicate its use in biomedical domains. Nanoceria were used by McGinnis and colleagues as

SOD mimetics to shield photoreceptor cells and avert light-induced retinal degeneration in rats.

The catalytic events are triggered by oxygen vacancies, created during the formation of nanoceria and flipped between Ce3+ and Ce4+ ions (Celardo et al. 2011). The SOD-like capabilities of nanoceria were confirmed through a competition analysis using ferricytochrome C and the formation of H_2O_2 as a byproduct. Mechanistic studies exhibited that a higher percentage of Ce3+ in the nanoceria resulted in greater catalytic activity (Celardo et al., 2011; Asati et al., 2009). Perez and colleagues found that polymer-coated nanoceria exhibited oxidase activity at low pH. Since H_2O_2 was not required for the process, they concluded that no or SOD activity was detected. Polymer-coated nanoceria exhibited universal oxidase activity (pH-dependent) toward their substrates like ABTS, DOPA, and TMB, and with maximum activity observed at pH 4.0. The oxidase-mimicking characteristics varied with pH, nanoceria size, and polymer coating thickness. The oxidation was enhanced by making the nanoceria core smaller and the polymer layer thinner.

An improved surface-to-volume ratio and a more porous, less dense coating, both of which allow for easier substrate molecule exchange (Asati et al., 2009; Zhang et al., 2016; Liu et al., 2017) were cited as the causes of the higher catalytic efficiency. The nanozyme's catalytic efficiency, low maintainenace, high stability, low cost production, and strong enzyme-mimicking activity of nanozymes are all among its many advantages. Nanozymes have substantial and obvious functions in the context of biosensing and immunoassay, as well as in the context of research-based analytical equipment, which is one reason why researchers from a variety of sectors are interested in creating and constructing nanozymes and carbon based nanomaterials (Bikash and Arup, 2021; Xin et al. 2019; Wong, 2021). Particularly accessible are nanomaterial-based synthetic enzymes, such as hybrid nanostructures and metal-organic frameworks (MOFs).

Nanozymes possess a distinctive catalytic property that enables them to enhance the detection sensitivity by amplifying the signals. In addition, it can be utilised to evaluate a wide range of analytes and biomarkers (Ali, 2004; Wei, 2013). The nanozyme offers substitute to enzyme-based immunoassay, i.e., Nanozyme-Based Enzyme-Linked Immunosorbent Assay (Nz-ELISA). An ultrasensitive ELISA was developed by researchers to detect the influenza A virus (Oh et al. 2018) using gold (Au) nanozyme and magnetic nanobead capture probes. Nanozymes are multifunctional and therefore, can be used in bio-signaling, diagnostics (neuroprotection, antiinflammation, chronic inflammatory diseases), and chemotherapeutic agents etc. (Das et al., 2007;

Oh et al., 2018). These nanozymes have the tendency to swamp the deficiencies of enzymes, such as storage difficulties, poor stability in harsh conditions, and high production cost.

2. Categories of Nanozymes

Nanozymes are broadly categorised into three categories: metal-based, metal oxide-based, pro-oxidant nanozymes, and carbon-based according to their composition and characteristics.

2.1. Metal Oxide-Based Nanozymes

2.1.1. Metal Oxide Nanoparticles (MNPs)

Metal oxides have numerous biomedical uses, counting target based drug delivery, biosensors, immunoassays, tissue regeneration, and adjuvants for cell separation etc. (Gupta et al., 2005; Gao et al., 2007). Metal oxide nanoparticles, typically considered chemically and physiologically inert, require a novel method for surface-based engineering and succeeding indent conjugation with functional molecules to be useful. Considering the recent finding of MNP's intrinsic catalytic property as a peroxidase mimetic, several publications have been published recently examining the new enzyme-like functions of metal oxide nanozymes (Gao et al., 2007).

2.2. Metal-Based Nanozymes

Metal-based nanozymes, like AuNPs and PtNPs, have catalytic activity such as oxidase, catalase, peroxidase, and SOD. An alloy of Fe_3O_4 with graphene (GO) with other nanoparticle systems, such as the Fe_3O_4-Pt Au-Pt and GO-Fe_3O_4-Pt nanocomposites have also been intensively studied. When metal-based nanozymes are coupled with other nanozymes, their synergistic activity greatly enhances catalytic activity (Liu et al. 2012; Kim et al. 2004).

2.3. Carbon-Based Nanozymes

Due to their unique enzyme-mimicking properties, Carbon-based nanozymes, mostly fullerene and GO, are keen tremendous interest (Song et al. 2010). Since they resemble both peroxidase and SOD, they are frequently utilized in biosensors and immunoassays to ramp up signals and detect analytes (Song et al. 2010; Guo et al. 2011).

2.4. Pro-Oxidant Nanozymes

The phrase "pro-oxidant nanozymes" refers to a situation in which a nanozyme causes free radicals to form in mammalian cells or suppresses the antioxidant system. Nanostructures with peroxidase and oxidase activity produces free radicals and are known as pro-oxidant nanozymes (Rahal et al. 2014; Liu, 2012; Wang et al. 2007).These reactions, similar to Fenton and Haber-Weiss, can also be carried out by transition metals like Iron and Copper, which generate high amount of free radicals (Wang et al. 2007; Rahal et al. 2014).

3. Applications of Nanozymes

Nanozyme's possess catalytic activity similar to that of enzymes but also exhibit nanomaterial properties such as fluorescence and photothermal properties, with superparamagnetic capabilities (Gupta et al. 2005). In addition, the physiochemical and catalytic properties of nanozymes have made them widely used for in vitro and in vivo detection, monitoring and treatment of disease (Gupta et al., 2005; Gao et al., 2007).

Nanozymes have already attracted much interest in recent years. However, they have many advantages over regular enzymes, including being long-lasting, stable, and affordable to produce at any scale (Gupta et al. 2005). Nanozymes have been exploited to detect various big and small molecules like DNA, protein and glucose/antibiotics (Bikash and Arup, 2021).

3.1. Nanozymes for Pathological Disease Diagnosis

Magnetic ferritin nanozyme (M-HFn) is comprised of nanocage conjugated with recombinant human ferritin (HFn) protein, and an iron oxide nanocore, used for tumor imaging and targeting (Fan et al. 2012). The HFn nanocage was found to help recognize malignant cells in addition to chemically binding to overexpressed transferrin receptor 1. Color responsiveness was enhanced using iron oxide nanocores and hydrogen peroxide in conjunction with one another. Using M-HFn nanozymes, researchers were able to identify between nine different types of cancer cells, with specificities ranging from 95% to 98%. Co_3O_4 nanozyme functionalized with an Avastin antibody can be discriminate between VEGF and other types of immunohistochemistry (Lu et al., 2022). It has been established that Co_3O_4 nanozyme is a viable alternative to naturally occurring enzymes in numerous applications. Meanwhile, peroxidase nanozyme-based staining techniques are being utilized to diagnose a variety of malignancies, including breast, hepatocellular, bladder, stomach, pancreatic, and colorectal cancers (Zhang et al., 2016; Hu et al., 2014).

3.2. Nanozymes for In-Vivo Imaging

Besides having enzyme-mimetic properties, nanozymes also have other notable physicochemical properties like as fluorescence, electricity, and paramagnetic characteristics. Nanozymes have also been extensively studied for their potential use in in-vivo disease monitoring and image processing, which take advantage of their unique physicochemical features. For example, the researchers used exceptional r2 relaxivity of Fe nanozymes to achieve cancerous cells in-vivo MRI after successfully imaging in-vitro tumor tissue. They used the peroxidase-mimetic activity of Fe-based nanozymes by administration of a single intravenous infusion of 125I-M-HFn NPs containing 11.2 grams of Fe, 500 grams of 125 I, and 45 grams of HFn (Ding et al., 2019; Peng et al., 2019).

3.3. Nanozymes for Multi-Drug Resistant Bacteria

Ions, molecules, and chemical compounds can all be distinguished using nanozymes in both qualitative and quantitative ways. Aside from destroying multidrug-resistant bacteria, they have also been utilized to break down organic pollutants. The use of antibacterial nanoparticles in treating bacterial illnesses is becoming increasingly popular (Chen et al., 2015). Although most nanomaterials may kill bacteria physically or chemically, this limits their effectiveness in treating multidrug-resistant pathogens. Multiple drug-resistant microorganisms have been eliminated, and other organic pollutants have been degraded with their help. Therefore, using antibacterial nanoparticles to treat bacterial infections is becoming increasingly fashionable. When dealing with multidrug-resistant microorganisms, the effectiveness of most nanomaterials is limited (Cobos, 2020; Zhu, 2020).

3.4. Metal-Organic Framework (MOF)-Based Nano-Enzymes and Their Biomedical Applications

As a replacement for regular enzymes, metal-organic frameworks (MOFs) and their generated products with enzyme-like activity could be used as antimicrobial medications, and in detecting and treating cancer. Their role as biosensing material, antibacterial compounds, cancer treatment, and similar simulation enzymes has been emphasized in recent studies (Wang et al., 2020; Chen, 2015; Singh, 2017; Fan et al. 2012).

3.5. Nanozymes in the Environmental Applications

Nanozymes are increasingly popular nowadays, due to their durability, high reactivity, and variety of sensing applications. In the future, nanozyme will be used in pollution management applications, for managing persistent organic pollutants which remain in the environment for an extended period. Such pollutants dispersed via wind, water, and animals in the food web. Pesticides, organophosphorus, insecticides, penetrants and disinfectants (bisphenol and chlorophenols) are the most frequent organic contaminants of water, soil, and food chain (Zhu et al., 2020; Barrios-Estrada et al., 2018). When it comes to pesticide monitoring and degradation, nanozymes have emerged in the last several years to help. Enzyme-like inhibition tests are used in some pesticide detection techniques (Liang, 2013; Singh, 2017; Zhu, 2022).

A schematic representation of multivariate applications of Nanozymes in Figure 1.

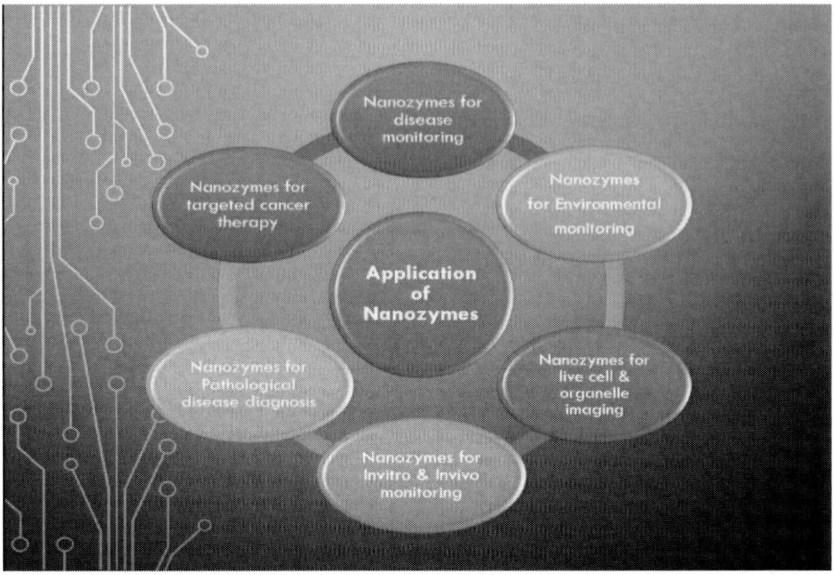

Figure 1. Multivariate applications of nanozymes in various domains.

3.6. Nanozymes Effective Solution for Analytical Methods

Nanozymes are more easily purified, functionalized, and modified with a capture probe than enzymes. Besides, nanozyme's length, surface modifications and geometry like nanorods, nanospheres, nanowires, nanosheets, etc., their catalytic activity can be controlled by varying morphology and size (Wang et al., 2019). Nanozymes can be replaced by traditional label moities and conjugated with a monoclonal antibody or nucleotide sequences (capture probe). This leads to amplification in signal then the traditional techniques (Zhu et al., 2017; Tao et al., 2020). Even more sensitive techniques with a lower detection limit can be developed with these bioconjugates for optical, electrochemical, or colorimetric assessments.

3.7. Nanozyme-Aided Catalytic Theranostics

The strategies and applications of nanozyme-based disease theranostics are the primary focus for the recent advances in Nanofield. In the near future opportunities and challenges of nanozyme-based catalytic theranostics will be a demanding thirst area of research (Zhang et al., 2019; Hu et al., 2014). By immediately boosting ROS via oxidase and peroxidase activity, nanozymes can be employed to kill malignancies because a lot of hydrogen peroxide builds up around tumour cells. Utilizing nitrogen-doped porous carbon nanospheres, Fan et al., 2018 created a nanozyme that exhibited oxidase, peroxidase, catalase, and SOD like activity. It catalysed the oxidation of O_2 and H_2O_2 in the presence of acid to form reactive oxygen radicals. Catalase and SOD can scavenge free radicals in a neutral environment. Targeting of tumour cells positioned in the lysosome was demonstrated using carbon nitrogen-modified ferritin nanoparticles which accelerated the production of highly toxic ROS to kill tumour cells. Ferritin-carbonitride nanozyme can considerably slow tumour growth, according to animal research. (Zhang et al., 2019; Hu et al., 2014).

3.8. Nanozymes towards Hypoxia Alleviation in Chemotherapeutic Agents

The catalase activity of nanozymes can reduce hypoxia and boost tumor's nutrition uptake. These techniques can increase the effectiveness of radiation, sonodynamic therapy, and photodynamic therapy while fostering tumor cell death. (Zhang et al., 2013). A tumor microenvironment contains more metabolites such as acid and hydrogen peroxide, than healthy tissues. Additionally, solid tumors are characterized by tumor hypoxia. It not only makes tumor cells more resistant to chemotherapy, but also serves as the primary catalyst for distant metastasis of the tumor.

A new nanozyme (PtFe@Fe_3O_4) responsive to tumor microenvironment (Li et al., 2019) has also been reported. PtFe@Fe_3O_4 can effectively combat the hypoxic environment of the tumour by decomposing hydrogen peroxide (H_2O_2) to produce O_2 and OH- radicals simultaneously to kill tumour cells. Fe_3O_4 serves as an electron pump to maintain Pt in the electron-rich state during the PtFe@Fe_3O_4 nanozyme catalysis process, increasing its catalytic activity. Due to the surface plasmon resonance (SPR), PtFe's catalytic activity was enhanced further when stimulated by near-infrared light. In an in-vivo treatment trial, fibre intervention was used to boost enzyme activity while the material's inherent photothermal properties were coupled to improve the effectiveness of the therapy of deep pancreatic cancer.

In order to combine tumour starvation and low-temperature photothermal therapy, Zhou et al. engineered nanozymes that bind exclusively to CD44-overexpressing tumour cells by loading glucose oxidase (GOx) into porous hollow nanoparticles of Prussian blue (PHPBNs). These were subsequently coated with hyaluronic acid (HA) via redox-cleavable coupling. In the hypoxic tumour microenvironment, PHPBNs's photothermal characteristics can destroy the tumour and prevent hypoxia-induced photothermal resistance by accelerating the breakdown of H_2O_2 to produce O_2. GOx-loaded PHPBNs can starve tumours, allowing for additional collaborative treatment. Additionally, this approach might therefore inhibit the formation of heat shock proteins (HSPs) following photothermal therapy to lessen the HSPs' susceptibility to PHPBN-mediated reduced photothermal therapy (Ruan et al. 2021).

Conclusion

Inorganic nanomaterials with enzyme-like capabilities process a catalytic function and exhibit distinguishing properties. Bright and ingenious ideas for nanoenzyme development, their engineering, and application highlight the future prospects of nanoenzymes (Zhu et al., 2017). These are used in detecting biomolecules and therapy of diseases caused by large molecules like chromosomes, peptides, organelles and small molecules like carbohydrates. Currently, nanozyme-based investigations and their applications are focused on novel biosensors for oncology diagnostics, pharmaceuticals, and environmental pollutants, and in immunohistochemistry (Singh et al., 2019). In addition, with the use of nanozymes in place of natural enzyme in cell-based assays, the cytological identification of cancer cells have improved a lot. This resulted in promising output in identifying of breast cancer cells (Li et al., 2019), human chronic myelogenous leukemia cells, melanoma tumor cells, cervical cancer cells, and squamous cancer cells (Wang et al., 2020).

The current roadblocks to the widespread usage of nanozymes in biotechnology, bioengineering, biosensors, embedded applications, immune-logic assays, immune reactions, and cancer diagnostics all necessitate geotechnical sustainability and investigations. (Zhang, 2008; Shin et al., 2015). Most nanozymes, except fullerene-based nanozymes, are not uniform in size and proportions. Therefore, the synthesis protocol must be improved to produce monodispersed nanozymes, because polydispersity of nanostructured material results in batch-to-batch variation in size and shape, and leads to changes in physicochemical properties. In the future, innovative computation-assisted technologies may make it possible to rationally build an atomically exact nanozyme for a specific activity using mathematical models.

References

Ali, S. S., Hardt, J. I., Quick, K. L., Sook Kim-Han, J., Erlanger, B. F., Huang, T., Epstein, C. J., & Dugan, L. L. (2004). A biologically effective fullerene (C60) derivative with superoxide dismutase mimetic properties. *Free Radical. Biol. Med,* 1191–1202.

Asati, A., Santra, S., Kaittanis, C., Nath, S. and Perez, J.M., Angew (2009). *Chem., Int. Ed,* 48, 2308–2312.

Barrios-Estrada, C., de Jesus Rostro-Alanis, M., Munoz-Gutierrez, B. D., Iqbal, H. M. N., Kannan, S., and Parra-Saldivar, R. (2018). "Emergent contaminants : Endocrine disruptors and their laccase-assisted degradation-A review." *Sci. Total Environ*, 612, 1516–1531.

Celardo, I., De, Nicola. M., Mandoli, C., et al. (2011). Ce (3)+ ions determine the redox-dependent anti-apoptotic effect of cerium oxide nanoparticles. *ACS Nano.* 5 (6) :4537–4549.

Chen, WY, Chang, HY, Lu, JK, Huang, YC, Harroun, SG., Tseng, YT, Li, YJ, Huang, CC, and Chang, HT (2015). "Self-assembly of antimicrobial peptides on gold nanodots : Against multidrug-resistant bacteria and wound-healing application." *Adv. Funct. Mater,* 25 : 7189–7199.

Cobos, M., De-La-Pinta, I., Quindös, G., Fernández, M. J., and Fernández, M. D. (2020). "Graphene oxide-silver nanoparticle nanohybrids : Synthesis, characterization, and antimicrobial properties." *Nanomaterials,* 10 : 376.

Das, M., Patil, S., Bhargava, N. et al. (2007). Auto-catalytic ceria nanoparticles offer neuroprotection to adult rat spinal cord neurons. *Biomaterials,* 28 (10):1918–1925.

Ding, H., Cai, Y., Gao, L., Liang, M., Miao, B., Wu, H., Liu, Y., Xie, N., Tang, A., Fan, K., Yan, X., & Nie, G. (2019)." Exosome-like nanozyme vesicles for H2O2-responsive catalytic photoacoustic imaging of xenograft nasopharyngeal carcinoma." *Nano Lett.* 19, 203–209. https://doi:10.1021/acs.nanolett.8b03709.

Dong, H., Fan, Y., Zhang, W., et al. (2019). Catalytic mechanisms of nanozymes and their applications in biomedicine. *Bioconjugate Chem,* 30 :1273–96.

Fan, K., Cao, C., Pan, Y., Lu, D., Yang, D., Feng, J., Song, L., Liang, M., & Yan, X. (2012). "Magnetoferritin nanoparticles for targeting and visualizing tumour tissues." *Nat. Nanotechnol,* 7, 459–464.

Fan, K., Xi, J., Fan, L., Wang, P., Zhu, C., Tang, Y., Xu, X., Liang, M., Jiang, B., Yan, X., & Gao, L. (2018). *In vivo* guiding nitrogen-doped carbon nanozyme for tumour catalytic therapy. *Nat. Commun.*, 1440. https://doi:10.1038/s41467-018-03903-8.

Feng, C, Bai, M, Cao, K, Zhao, Y, Wei, J, and Zhao Y (2017). "Fabricating MnO_2 Nanozymes as Intracellular Catalytic DNA Circuit Generators for Versatile Imaging of Base Excision Repair in Living Cells." *Advanced Functional Materials,* 27, 45 :1702748.

Gao, L., Zhuang, J., Nie, L., Zhang, J., Zhang, Y., Gu, N., Wang, T., Feng, J., Yang, D., Perrett, S., & Yan, X. (2007). "Intrinsic peroxidase-like activity of ferromagnetic nanoparticles." *Nature Nanotechnology,* 577-583.

Ge, L., Li, Q., Wang, M., Ouyang, J., Li X., and Xing, M.M. (2014). Nanosilver particles in medical applications : synthesis, performance, and toxicity. *Int. J. Nanomedicine*, 9 p.2399.

Guo, Y., J. Li., and S. Dong. (2011). "Hemin functionalized graphene nanosheets-based dual biosensor platforms for hydrogen peroxide and glucose," *Sensors and Actuators B: Chemical*, vol. 160, no. 1, pp. 295–300.

Gupta, A. K., and Gupta, M. (2005). "Synthesis and surface engineering of iron oxide nanoparticles for biomedical applications," *Biomaterials*, 3995–4021.

Hu, D., Sheng, Z., Fang, S., Wang, Y., Gao, D., Zhang, P., et al. (2014). Folate receptor-targeting gold nanoclusters as fluorescence enzyme mimetic nanoprobes for tumor molecular colocalization diagnosis. *Theranostics*, 4, 142–153.https://doi:10.7150/thno.7266.

Jena, BR, and Chakraborty, A, (2021). "Recent Advancement of Carbon-based Nanomaterials (CBNs)." *Acta Scientific Pharmaceutical Sciences*, 01-02.

Li, J., Liu, M., Jiang, J., Liu, B., Tong, H., Xu, Z., Yang, C., & Qian, D. (2019). "Morphology-controlled Electrochemical Sensing Properties of CuS Crystals for Tartrazine and sunset Yellow." *Sensors Actuators B: Chem,* 288, 552–563.

Li, S, Shang, L, Xu, B, Wang, S, et al., (2019). A Nanozyme with Photo-Enhanced Dual Enzyme-Like Activities for Deep Pancreatic Cancer Therapy. *Angew. Chem., Int. Ed. Engl.*, 58, 12624 —12631.

Liang, M., Fan, K., Pan, Y., Jiang, H., Wang, F., Yang, D., Lu, D., Feng, J., Zhao, J., Yang, L., & Yan, X. (2013). "Fe_3O_4 magnetic nanoparticle peroxidase mimetic-based colorimetric assay for the rapid detection of organophosphorus pesticide and nerve agent." *Anal. Chem*, 85, 308–312.

Liu, J., Hu, X., Hou, S., Wen, T., Liu, W., Zhu, X., Yin, J. J., & Wu, X. (2012). "Au@Pt core/shell nanorods with peroxidase- and ascorbate oxidase-like activities for improved detection of glucose," *Sensors and Actuators B : Chemical*, 166-167, 708–714. https://doi:10.1016/j.snb.2012.03.045.

Liu, Y., Ai, K., Ji, X., Askhatova, D., Du, R., Lu, L., and Shi, J. (2017). *J. Am. Chem. Soc,* 139, 856–862.

Lu, G., Wang, X., Li, F. et al. (2022). Engineered biomimetic nanoparticles achieve targeted delivery and efficient metabolism-based synergistic therapy against glioblastoma. *Nat Commun* 13, 4214 https://doi.org/10.1038/s41467-022-31799-y.

Oh, S., Kim, J., Tran V. T. et al. (2018). "Magnetic nanozyme-linked immunosorbent assay for ultrasensitive influenza A virus detection," *ACS Applied Materials & Interfaces*, vol. 10, no.15, pp. 12534–12543.

Pankhurst, Q.A., Connolly, J., Jones S.K., and Dobson, J., J. Phys. D. (2003) Applications of magnetic nanoparticles in biomedicine. *Appl. Phys.,* 36 p. R167.10.1088/0022-3727/36/13/201.

Peng, C., Hua, M. Y., Li, N. S., Hsu, Y. P., Chen, Y. T., Chuang, C. K., Pang, S. T., & Yang, H. W. (2019). "A colorimetric immunosensor based on self-linkable dual-nanozyme for ultrasensitive bladder cancer diagnosis and prognosis monitoring." *Biosens. Bioelectron,* 581–589.

Rahal, A., Kuma, R. A., Singh, V., Yadav, B., Tiwar, R., Chakraborty, S., and Dhama, K. (2014). Oxidative stress, prooxidants, and antioxidants : the interplay. *Biomed Res Int.,* 761264. https://doi:10.1155/2014/761264.

Ruan C, Su K, Zhao D, Lu A, Zhong C. Nanomaterials for Tumor Hypoxia Relief to Improve the Efficacy of ROS-Generated Cancer Therapy. Front Chem. 2021 Mar 19; 9:649158. doi: 10.3389/fchem.2021.649158. PMID: 33954158; PMCID: PMC8089-386

Shin, Ho Yun., Tae, J Park., and Moon, il Kim. (2015). "Recent Research Trends and Future Prospects in Nanozymes." *Journal of Nanomaterials*. https://doi.org/10.1155/2015/756278.

Singh, S. (2019). "Nanomaterials Exhibiting Enzyme-like Properties (Nanozymes): Current Advances and Future Perspectives." *Frontiers in Chemistry,* 7 (FEB). https://doi.org/10.3389/FCHEM.2019.00046/FULL.

Singh, S., Tripathi, P., Kumar, N., and Nara, S. (2017). "Colorimetric sensing of malathion using palladium-gold bimetallic nanozyme." *Biosens. Bioelectron*, 92, 280–286.

Song, Y., X., Wang, C. Zhao, K. Qu, J. Ren, and X. Qu. (2010). "Labelfree colorimetric detection of single nucleotide polymorphism by using single-walled carbon nanotube intrinsic peroxidaselike activity," *Chemistry—A European Journal*, vol. 16, no. 12, pp. 3617–3621.

Sun, C., J.S. Lee. and Zhang, M. (2008). Magnetic nanoparticles in MR imaging and drug delivery *Adv. Drug Delivery Rev.* 60 p.1252.10.1016/j.addr.2008.03.018.

Sun, Y., Zheng, L., Yang, Y., Qian, X., Fu, T., Li, X., Yang, Z., Yan, H., Cui, C., & Tan, W. (2020). "Metal–Organic Framework Nanocarriers for Drug Delivery in Biomedical Applications." *Nano-Micro Letters*, 12, https://doi:10.1007/s40820-020-00423-3.

Tao, X., Wang, X., Liu, B., and Liu, J. 2020. "Conjugation of Antibodies and Aptamers on Nanozymes for Developing Biosensors." *Biosens. Bioelectron.,* 168, 112537.

Tartaj, P., Morales, M.D., Verdaguer, S.V.; T. Gonzalez-Carreno and C.J. Serna. (2003). The preparation of magnetic nanoparticles for applications in biomedicine. *J. Phys. D. Appl. Phys.* 36 p. R182.10.1088/0022-3727/36/13/202.

Wang, C., Wang, H., Xu, B., et al. (2020). Photo-responsive nanozymes : mechanism, activity regulation, and biomedical applications. *View.,* 2 :20200045.

Wang, H., Wan, K., and Shi, X. (2019). "Recent Advances in Nanozyme Research." *Adv. Mater.* 31 (45), 1805368.

Wang, W., Li, B., Yang, H., Lin, Z., Chen, L., Li, Z., Ge, J., Zhang, T., Xia, H., Li, L., & Lu, Y. (2020). Efficient elimination of multidrug-resistant bacteria using copper sulfide nanozymes anchored to graphene oxide nanosheets. *Nano Research*, 13(8), 2156–2164. https://doi.org/10.1007/s12274-020-2824-7.

Wang, X., Song, X., Si, L., Xu, L., and Xu, Z. (2020). "A Novel Biomimetic Immunoassay Method Based on Pt Nanozyme and Molecularly Imprinted Polymer for the Detection of Histamine in Foods." *Food Agric. Immunol.* 31 (1), 1036–1050.

Wang, Z. Y, X., Zhang, X. Xiao, Chang, C. K., and Xu. B. (2007). "A supramolecular-hydrogel-encapsulated hemin as an artificial enzyme to mimic peroxidase," *Angewandte Chemie Inter-national Edition*, 4285–4289.

Wei, H. and E. Wang. (2013)."Nanomaterials with enzyme-like characteristics (nanozymes) : next-generation artificial enzymes," *Chemical Society Reviews*, 6060–6093.

Wong, E. L. S., Vuong, K. Q., and Chow, E. (2021). Nanozymes for Environmental Pollutant Monitoring and Remediation. *Sensors,* 21,408.

Xin L, Wang, L, Du, D, Ni, L, Pan., J, and Niu, X, (2019). "Emerging applications of nanozymes in environmental analysis: opportunities and trends." *TrAC Trends in Analytical Chemistry,* 120:115653.

Zhang, D, Zhao, Y-X., Gao, Y-J, Gao, F-P, Fan, Y-S, Li, X-J, Duan, Z-Y, and Wang, H, (2013). A Review of One-dimensional TiO2 Nanostructured Materials for Environmental and Energy Applications. *J. Mater. Chem. B*, 1, 1 —8.

Zhang, J., Zhuang, J., Gao, L., Zhang, Y., Gu, N., Feng, J., Yang, D., Zhu, J., & Yan, X. (2008). "Decomposing phenol by the hidden talent of ferromagnetic nanoparticles," *Chemosphere*, 73 : 1524–1528.

Zhang, T., Cao, C., Tang, X., Cai, Y., Yang, C., and Pan, Y. (2016). "Enhanced peroxidase activity and tumour tissue visualization by cobalt-doped magnetoferritin nanoparticles." *Nanotechnology*, 045704.

Zhang, W., Hu, S., Yin, J.J., He, W., Lu, W., Ma, M., Gu, N. and Zhang, Y. (2016). Prussian Blue Nanoparticles as Multienzyme Mimetics and Reactive Oxygen Species Scavengers. *J. Am. Chem. Soc*, 138, 5860–5865.

Zhang, Y, Jin Y, Cui H, Yan X, Fan K. (2019). Nanozyme-based catalytic theranostics. *RSC Adv.*, Dec 23;10(1):10-20. https://doi:10.1039/c9ra09021e.

Zhou, D., Fang, T., Lu, Lq. et al. (2016). Neuroprotective potential of cerium oxide nanoparticles for focal cerebral ischemic stroke. *J. Huazhong Univ. Sci. Technol.* 36, 480–486 https://doi.org/10.1007/s11596-016-1612-9.

Zhu, X., Mao, X., Wang, Z., Feng, C., Chen, G., and Li, G. (2017). "Fabrication of Nanozyme@DNA Hydrogel and its Application in Biomedical Analysis." *Nano Research.* 10, 959–970.

Zhu, Y. Y., Wu, J. J. X., Han, L. J., Wang, X. Y., Li, W., Guo, H. C., and Wei, H. (2020). "Nanozyme sensor arrays based on heteroatom-doped graphene for detecting pesticides." *Anal. Chem*, 92, 7444–7452.

Index

C

D

M

N

O

P

Q

R

S

Editors' Contact Information

Dr. Seema Nara
Associate Professor
Department of Biotechnology
Motilal Nehru National Institute of Technology Allahabad,
Prayagraj, Uttar Pradesh, India
seemanara@mnnit.ac.in

Ms. Smriti Singh
PhD Scholar
Department of Biotechnology
Motilal Nehru National Institute of Technology Allahabad,
Prayagraj, Uttar Pradesh, India